# MAPPING AND IMAGINATION
## IN THE GREAT BASIN

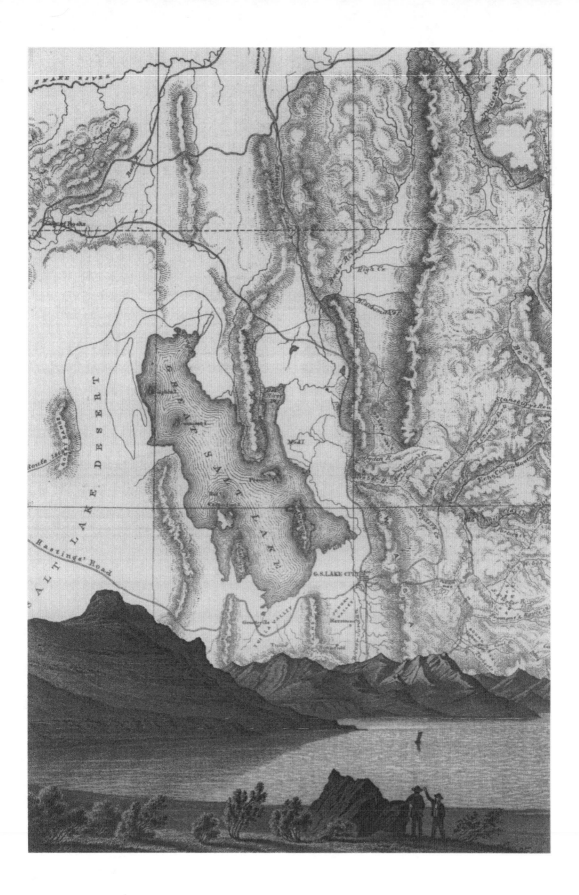

# MAPPING
# AND IMAGINATION
# IN THE GREAT BASIN

*A Cartographic History*

## RICHARD V. FRANCAVIGLIA

UNIVERSITY OF NEVADA PRESS

*Reno & Las Vegas*

www.unpress.nevada.edu

University of Nevada Press, Reno, Nevada 89557 USA

Copyright © 2005 by University of Nevada Press

Photographs copyright © 2005 by author

All rights reserved

Manufactured in the United States of America

Design by Barbara Jellow

Library of Congress Cataloging-in-Publication Data

Francaviglia, Richard V.

    Mapping and imagination in the Great Basin : a cartographic history /
Richard V. Francaviglia.—1st ed.

      p. cm.

    Includes bibliographical references and index.

    ISBN 978-0-87417-609-4 (hardcover : alk. paper)—ISBN 978-0-87417-617-9
(pbk. : alk. paper)

    1. Cartography—Great Basin Region—History. 2. Great Basin—
Discovery and exploration. 3. Great Basin—Geography. I. title.

    GA102.6F3 2004

    912.79—DC22

                            2004023023

*This book has been reproduced as a digital reprint.*

*Frontispiece:* ENVISIONING GREAT SALT LAKE IN THE 1850s—a
composite image of the landscape scene "WEST END OF FRÉMONT'S
I[SLAND], AND PROMONTORY RANGE, LOOKING NORTH. G.S.
LAKE" from Howard Stansbury's *Exploration of the Valley of the Great Salt
Lake* (1852) and a portion of the "Map of Wagon Routes in Utah
Territory" in Captain J. H. Simpson's *Report of Explorations Across the Great
Basin of the Territory of Utah* (1859). Author's collection.

*Dedicated*
*To*
*Jenkins and Virginia Garrett,*
*lovers of history and maps*

# Contents

# Illustrations

# INTRODUCTION

# Maps and Meaning

ALTHOUGH TRAVELERS MOVING THROUGH Salt Lake City's busy international airport are usually concerned about security screening delays and catching their flights on time, they are also presented with a number of displays and images that highlight the city's natural setting. Without even gazing out the airport's windows, the traveler glimpses images of the deserts, mountains, and inland lakes that make this city's environment so distinctive. These images appear on posters and displays in the airport's shops and concourses. One in particular—a stunning thirty-foot-wide mural by A. C. Bliss in the airport's concourse (figure 0.1)—is especially revealing. Called *The Discoverers,* the mural offers a glimpse westward from the Wasatch Mountains down into the sprawling Salt Lake Valley. In addition to the mural's mysterious see-through, line-drawn human figures, two other elements draw viewers' attention: The mountains themselves are rugged and defined by jagged, vertically oriented lines, while the landscape of the adjacent lowlands is divided horizontally into a series of geometric forms bounded by straight lines.

The jaded traveler may at first consider this mural to be civic art with a message—another attempt to promote the city. But it is much more. Looked at more closely, and in historical perspective, *The Discoverers* is part of a rich tradition of visually portraying the Great Basin. It uses much the same vantage point that artists used in depicting the Mormons' arrival in the Salt Lake Valley in 1847. *The Discoverers* thus fits into a long tradition of art history, but it also has a cartographic or maplike quality: it conveys something about the design of the valley's geographic setting, and also something about the artist's vision of how nature and humanity are spatially arranged here.[1] If *The Discoverers* had been drawn or painted from a slightly higher perspective, it would qualify as a bird's-eye view, a type of map that provides the map reader a vista from several hundred feet aloft.[2] Bird's-eye views have expanded our definition of maps, which many people still think must be drawn planimetrically (that is, as if looking straight down toward the mapped location) but may in fact be drawn from any elevated perspective.

*The Discoverers* not only hints at the power of the human mind to define and organize space; it also depicts that layout and thus serves as a map in the broadest sense of the word. This mural-as-map is ultimately cartographic. By cartographic, I refer to the human propensity to perceive order in the landscape, then reflect on that order and depict it in illustrations of all kinds, including traditional maps. Broadly defined, a map is any device that depicts spatial relationships. We can use maps to decipher many things in addition to places: for example, to map genomes in order to determine the genetic makeup of an organism; or map the circulatory system of the human body. Mostly, however, we understand a map to tell us about place, or places.

Consider the great variety of place-oriented maps that we experience today. Once on the aircraft, the traveler leaving Salt Lake City (or any other airport, for that matter) may see the plane's route depicted either on a drop-down TV monitor that shows the flight's progress, or in the form of a paper map showing the airline's routes at the back of the flight magazine. Looking at such maps, the traveler flying westward from Salt Lake City will note that his or her airplane soon reaches a large oval-shaped area—a region that is featured on maps in ways that reveal something about its character by suggesting its openness and ruggedness: shaded in earth tones, surrounded by somber mountains, and nearly devoid of cities. This virtually blank area is the Great Basin, and the maps that the traveler sees of it are only the latest in a long series of cartographic products that have depicted—or attempted to depict—the region in ways that reveal something about its character. This book is generally about the process by which maps and related images reveal the character of places. More particularly, it is about how mapmakers have depicted the Great Basin in the tradition of Western, which is to say European and European American, mapmaking. It is also about the

FIGURE I.I.

*The Discoverers* mural by A. C. Bliss (1996), Salt Lake City International Airport.
2001 photo by author

people who make those maps and the other people who motivate them to do so.

The Great Basin is not only part of, but actually the heart of, the great American West. In *Western Places, Western Myths,* geographer Gary Hausladen urges us to recall that "the evolution of the American West is a continuous process, and the delimitation of the region and the understanding of the varied components of the process change over time."[3] An important part of this process of visualizing the West and its subregions is cartographic. Imagine, for a moment, that we can be privileged enough to place dozens of maps of the Great Basin from over a long period of time—say several hundred years—side by side for comparison. This is exactly what I shall treat the reader to in this book. Looking at these maps carefully will suggest a process of change or evolution. Some of the early maps identify this region only as terra incognita—totally unknown land. By consulting maps of this region stretching back several centuries, we shall see that *terra incognita* is a relative term that is dependent on culture and time. Starting with maps from the 1500s, we shall see the region take shape in fits and starts timed to the pulse of colonial exploration and territorial expansion. It did so through individual maps that found their way into the hands of the elite via a growing network of map and book distributors.

Although this book is about historic maps and the real places they portray, I begin it with the assumption that most readers will not be cartographic historians, nor with they be very familiar with the geography of the Great Basin. If readers wish to know more about the Great Basin's unique physical environment, I suggest two informative books—*The Desert's Past* by Donald Grayson and *The Sagebrush Ocean* by Stephen Trimble.[4] Then, too, readers may wish to read William Fox's *The Void, the Grid, and the Sign*.[5] In that provocative book, Fox suggests that the Great Basin is such an overwhelming region that people were ultimately compelled to configure its surface into recognizable shapes as a way of overcoming its vastness. To paraphrase Fox, people have actually made the region itself into a map in order to get a grip on it. In *Believing in Place,* I suggested that the region's landscapes are so potent that they have affected deep human beliefs—the spiritual beliefs of both Native Americans (notably Paiute and Shoshone) and Anglo-Americans (notably including Mormons, miners, and other settlers).[6] All of the books mentioned above suggest that the region is not only a real place, but that the human imagination plays an especially strong role in shaping that place.

Maps can help us understand the way in which place becomes recognized as either familiar or exotic. Because this book is about the process by which the region's geography was comprehended through maps and other visual images, I must also make brief reference to several important cartographic history books that precede it. These include overviews on the subject, such as Phillip Allen's *Mapmaker's Art,* John Goss's *The Mapping of North America,* and E. W. Gilbert's *The Exploration of Western America*.[7] I also refer readers to philosophical works about cartography, including Denis Wood's short but provocative *The Power of Maps* and the late J. B. Harley's *The New Nature of Maps*.[8] These books contain a wealth of historic map images as well as information on how to read maps critically. Readers are also invited to go online and explore historic map collections such as those in the Library of Congress, or in Spanish or British archives and universities (like the informative "Cartographic Connections" website at the University of Texas at Arlington). Then, too, they might visit the websites of private collectors, such as David Rumsey, whose extensive map collection focuses on the Americas.[9] A number of his maps appear in this book. Through the Internet, Rumsey's superb personal cartographic history library in San Francisco can now be viewed anywhere on earth.

Cartography is the art and science of mapmaking. The history of cartography is also something of an art and science, as many maps come to us through diligent searches in the arts, humanities, and sciences. As a cartographic historian, I owe a great deal to scholars and map collectors. As author, I also owe a personal debt to many people who assisted me with my research. These include Peter Blodgett of the Huntington Library; Ben Huseman and Cammie Vitale Shuman of the DeGolyer Library and Toni Nolen of the Fondren Library at Southern Methodist University in Dallas; Erik Carlson and Paul Oelkrug of the Special Collections Division of the University of Texas at Dallas; Philip Notarianni, director, Utah State Historical Society; Professor Paul Starrs, Geography Department, University of Nevada at Reno; Joyce M. Cox, head of Reference Services, Nevada State Library and Archives in Carson City; Michael N. Landon, Ronald G. Watt, Brent M. Reber, and William Slaughter of the Archives, Church of Jesus Christ of Latter-day Saints, Salt Lake City; Emery Miller of WorldSat International; and Ingo Schwarz of the Alexander von Humboldt Forschungsstelle, Berlin Brandenburgische Akademie der Wissenschaften. Professional TV weather forecaster and map collector David Finfrock of Cedar Hill, Texas, shared maps from his personal collection. David Rumsey did likewise with his map collection in San Francisco. David Myrick of Santa Barbara shared information about the railroads of the Great Basin. At my university—the University of Texas at Arlington—many people helped with this project: cartographic archivist Kit Goodwin and university archivist Gary Spurr of the Special Collections Division of the Libraries helped me locate, and provided me copies of, maps from this extensive archive; cartographic historian Dennis Reinhartz shared insights about Spanish maps of North America; medieval historian Bede Lackner and French historian Steve Reinhardt assisted by translating some of the Latin and French on sixteenth-century and seventeenth-century maps; friend and former UTA student Nancy Grace kindly provided me access to a copy of her original 1854 map of the United States that so graphically depicts the Great Basin; Gerald Saxon, dean of UTA's Libraries, provided valuable input on early-nineteenth-century maps of the region; my colleague David Buisseret also shared his extensive knowledge of maps of discovery when I presented the outline of this book as a paper at the 2001 Society of the History of Discoveries meeting in Denver; History Department chair Don Kyle encouraged this project at every turn; and lastly, I must acknowledge the steadfast support of Ann Jennings, administrative secretary at UTA's Center for Greater Southwestern Studies and the History of Cartography. She typed innumerable drafts of this book and provided encouragement for me to keep writing during times of dark distraction that followed the events of September 11, 2001, reminding me that history shines light even as it casts shadows.

MAPPING AND IMAGINATION
IN THE GREAT BASIN

# I

## Comprehending the Great Basin

TO WAYFARERS IN THE EARLY twenty-first century, the Great Basin is one of those seemingly empty spaces that once punished the traveler but are now easily crossed unless one makes a mistake or miscalculation. Moving miles high above the region in an airplane, passengers who bother to look down see a series of dark, rugged mountain ranges that alternate with white salt flats. For them, this countryside rolls by in a little more than an hour and a half as their plane crosses the entire region between the Wasatch Range near Salt Lake City and the Sierra Nevada just west of Reno. That trip goes quickly enough, but crossing the Great Basin in an automobile, even with a good road map by your side to help orient you and explain some of the mysteries of the place, is a bit more of an adventure. With its forlorn-sounding names like Elko, Winnemucca, and Battle Mountain, the Great Basin is somewhat daunting to this day. Even at seventy-five or eighty miles per hour, crossing the region requires a full day's drive through some of the most sparsely populated countryside in America. Although dotted by a few towns, the well-engineered Interstate 80 runs through wild, wide-open spaces. Try Highway 50, which parallels the interstate about a hundred miles south, and you will have even more solitude on what Nevadans proudly call "the loneliest road in America." Even so, with today's CD players and air-conditioning, this journey will be a far cry from what explorers and travelers experienced just two centuries ago. Still, many modern-day travelers are humbled by the region's landscape of wide-open spaces. As one recent traveler put it, "I didn't know there was still so much empty space left anywhere in the United States."

In this chapter, I shall sketch out the basic outlines of Great Basin geography, emphasizing the physical features that make the region so distinctive. Before doing so, however, I would like to discuss—perhaps "deconstruct" might be a better word—the base map that I prepared for this introduction (figure 1.1). This map is intended to acquaint you with the Great Basin. Like all maps, it embodies a degree of personal style, but that must be subservient to the map's main purpose—to communicate relevant information about a place in ways that you, the map reader, will understand. Like all mapmakers, I was restricted by technology (the printed medium), format (in this case, the size of a page), cost (color was prohibitively expensive), and purpose (to inform, but not overwhelm, the reader). Like all maps, the base map thus represents a number of compromises. It can only show so much, and therefore what is depicted is the result of a conscious editing process that began well before I ever put pen to paper.

In order to communicate effectively, the map—like any representation of a place—also had to adhere to several conventions that are easy to overlook simply because we take them for granted:

*Orientation.* The map is oriented with north at the top. I could have drawn it with south at the top, but that would have confused readers who are so used to north being "up" that they would have considered the map to be "upside down."

*Placement.* Map readers automatically assume that the map accurately depicts the positions of individual places—mountains, rivers, cities—in their geographically correct positions. This seems natural but is actually cultural. For most of human history, maps relied on relative rather than exactly pinpointed positions.

*Proportion.* Map readers in our cultural tradition need to know how much territory is covered, and they assume that the proportions of the map are consistent throughout. This, too, is cultural in that most maps before the Age of Exploration were relative; even today, people drawing maps will often note that they are "not to scale" if the distances and proportions within the map vary.

*Projection.* The base map is produced on a flat piece of paper but attempts to reproduce a fairly large piece of the earth's surface that actually curves.

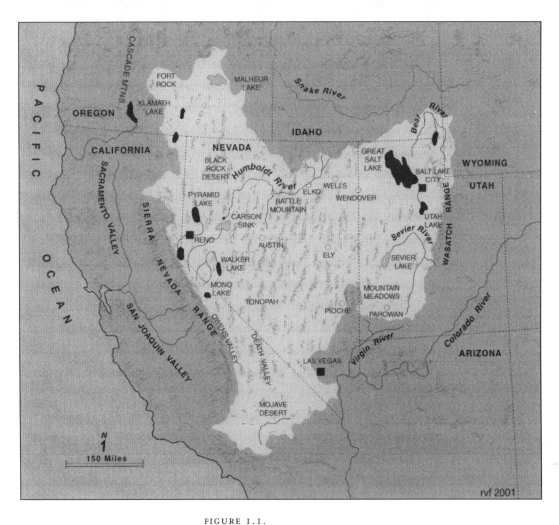

FIGURE I.I.

Base Map of the Great Basin (map by author)

This required a certain amount of distortion and involved yet another compromise.

*Content.* Maps are always intended to serve particular purposes. To acquaint the reader with the Great Basin, I had to greatly simplify the map's content. For example, only the major rivers and lakes are shown, and the mountains are depicted in a very stylized way to give the reader the general impression of the region's topography. Showing every geographic feature would have been impossible. Every cartographer faces this challenge and must be selective in representing places.

*Symbols.* The base map uses a number of symbols that are commonly understood in our culture—dotted lines for political boundaries, single lines for rivers, circles for towns and cities, hachured lines for mountain slopes. These seem natural enough, but they are really abstract symbols that map readers in our culture come to understand only through repeated experience and exposure. Every mapmaker faces the challenge of using understood symbols. Historically, people have used widely varied symbols for similar features—for example, a glyph for a mountain—that would be unintelligible to us unless we were instructed in their meaning.

*Authorship.* As my initials at its bottom edge claim, I drafted this base map. However, like all cartographers, I relied on information from other sources, including other maps, to guide me. I then crafted that information into a seemingly new image, a unique map unlike any other. Yet, in so doing, I depended on others for both inspiration and information. Those maps provided the basic parameters—political boundaries, natural features, communities—that I used to create the base map. Moreover, although I conceived of the basic map design, and sketched out a prototype, I had help, as Joel Quintans transcribed all of the basic lines and parameters using a computer program.[1] This confession serves as a reminder that maps are always collaborative efforts, regardless of how strongly individual mapmakers may assert otherwise.

Now that I have shown just how culturally relative this base map is—and all maps are—I would like to discuss the stunning geographic environment that it depicts. Note that I shaded the areas outside of the Great Basin as a way of drawing your attention to what lies *within* the region. Sequestered from surrounding areas by prominent mountain ranges on east and west lies a landscape of desert basins and mountains. I like the way the region's whiteness on the map makes it jump out, making a claim to openness, perhaps even suggesting emptiness, but also throwing "light" onto it. The base map is meant to convey the sense that this region is distinctive from what lies outside of it, and it is to that subject—the region's physical environment—that we now return.

From a European American perspective, the Great Basin is a harsh, sparsely populated desert landscape. It occupies a huge area, 165,000 square miles of Interior Western North America. Like all deserts, it simultaneously stimulates the imagination with thoughts of adventure and cautions us to comprehend its dangers. Seen in one of its harsher moods—say, on a blazing hot summer day or during an early spring sandstorm—the Great Basin seems an infernal, unforgiving place to be avoided. Consider travelers crossing this largely desert region a century and a half ago who faced challenges and privations mile after mile. The words of one traveler crossing a stretch of the Great Basin in 1850 pretty much summarize popular opinion. Dan Carpenter, who traversed this area on the way to California's gold rush, put it as succinctly as any European American ever has: "This is the poorest and most worthless country that man ever saw." Such desolation called for answers, but Carpenter felt that the landscape here was so irredeemable that even God had given up on it: "No man that never saw [it] has any idea what kind of a barren, worthless, valueless, d—d mean God forsaken country" awaits the traveler here. But Carpenter quickly corrected himself, concluding that the Great Basin was, in fact, "not God forsaken for He never had anything to do with it."[2]

In 1870, British traveler/writer William Bell crossed the Great Basin on the newly completed Central Pacific line. Bell stated that the distance from the Wasatch to the Sierra Nevada was 721 miles by railroad. Even by rail, it took a couple of days to cross this huge region. "It is a vast desert," Bell wrote, "considerably larger than France." An astute observer of physical geography, Bell succinctly characterized the Great Basin as "covered with short volcanic mountain ranges; it possesses a fertile soil, but suffers from an insufficient rain-fall; none of its scanty streams enter the sea, but each discharges its waters into a little lake and remains shut up within its own independent basin."[3]

Imagine a land so seemingly bereft of water that its few rivers were oddities in themselves. Their very presence and their peculiar behavior often invited comment. All of the region's rivers originated outside it, and in fact never reached the sea. The major river valley along which Carpenter traveled seemed a mockery: the Humboldt River flowed several hundred miles before quietly, even inconspicuously, vanishing into the sands. These rivers were etched into an equally enigmatic landscape, so arid that its plants grew widely spaced to conserve moisture, leaving bare patches between them. In some places, the soil was so saline that plants could not grow at all. However, in this same land, paradoxically, are several substantial lakes sustained only by rainwater or melting snows from the adjacent mountains; and throughout the region, many dry lakebeds mock the explorer with mirages of water that dance on shimmering heat waves. The Interior West, especially the Great

Basin, has only a few counterparts, and those that do qualify are equally austere—for example, depressed portions of the North African Sahara Desert, the "Dead Heart" of central Australia, and the Mongolian/ Gobi Desert of interior Asia. These analogous areas, not coincidentally, likewise remained largely unexplored, or at least uncharted, by Westerners until the 1800s.

However, imagine another side of the same coin—the same region seen either in a more sanguine mood or by someone who has known no other place. To the Indians of this region, including the Shoshone, Paiute, and Washo, the region offered sustenance—provided one knew where, when, and how to look for it. To these indigenous peoples, the region offered plants and animals to eat, including that seeming enigma in the desert—fish. Surviving here depends on how well one knows the environment. To hunters and gatherers (which pretty much describes the region's peoples before the introduction of agriculture), the Great Basin offered a bounty of resources that were tapped seasonally. Doing so required considerable mobility, for to stay in one particular place very long meant overstaying one's welcome. But if adjusting to this environment required periodic movement, certain bands or tribes of native peoples were not constantly in motion. They recognized that some places, notably near freshwater lakes, offered semipermanent retreats. These they called home for extended periods. And yet, when the time or season was right, when the rains moved and the ground warmed just so, the native people would move toward the new opportunities provided elsewhere.

The geographic design of the Great Basin impressed traveler and mapmaker alike. Consider first the hydrology. It is a region of *interior drainage* into numerous closed basins which may contain (a) *lakes* (either fresh or salty), (b) intermittently *marshy areas* or *seasonal lakes,* or (c) *playas* that are perennially dry except immediately after rare rainstorms that may be separated by a year or more of intense drought. This means that all precipitation that falls here either evaporates or remains. The Great Basin challenges many assumptions about hydrology, but none more than the popular one that rivers ultimately reach the sea. The song "Unchained Melody" (1955) asserts that "lonely rivers flow to the sea, to the sea, / to the open arms of the sea," as if this is inevitable. In the popular song "I Can't Help Falling in Love" (1960) Elvis Presley claimed, "Like a river flows, surely to the sea," his love for a woman was destined to be.

FIGURE I.2

Walker Lake, Nevada, looking southwestward toward the Sierra Nevada.
April 2000 photo by author

The river-to-sea metaphor is still a potent undercurrent in popular psychology: In recounting her experiences aboard a navy ship bound for the Persian Gulf, Boatswain's Mate 3d Class Laura E. Dwyer felt that she was "part of the force that was going to war, bound for it, destined for it like a river that runs into the sea."[4] These sentiments also characterized those in the nineteenth century who first encountered the Great Basin and were mystified by its hydrology. However, since the 1840s, observers who experienced the Great Basin firsthand warned others about its enigmatic rivers, which do not ever reach the sea. By the 1850s, travelers frequently reported that rivers here simply sank into the sands. In the early 1860s, Mark Twain joined the chorus of those trying to set the record straight. The Carson River, he stated, empties into "the 'Sink' of the Carson, a shallow melancholy sheet of water some eighty or a hundred miles in circumference." The river here, Twain noted, "is lost—sinks mysteriously into the earth and never appears in the light of the sun again—for the lake has no outlet whatsoever." Twain observed that these disappearing rivers were commonplace in this peculiar region. "There are several rivers in Nevada," he wrote, "and they all have this mysterious fate." They terminate in sinks or lakes like "Carson Lake, Humboldt Lake, Walker Lake, Mono Lake, [which] are all great sheets of water without any visible outlet." Twain marveled that these lakes "remain always level full, neither receding nor overflowing" (fig. 1.2).

What caused this phenomenon? Twain humorously observed, "What they do with their surplus is only known to the Creator."[5] We now know that these lakes are in a delicate balancing act. They result from rainwater and melting snows coursing down from the normally well-watered mountains. In the valleys, the lakes are subject to evaporation and percolation into subsurface aquifers. Lake levels can fluctuate, but Twain's frame of reference was short enough that, to him, such lakes seemed, as he put it, "permanent." As residents of California's Owens Valley learned long ago, diversion of river waters can result in a sapphirelike Owens Lake drying up into a salt flat as the city of Los Angeles captured its source. Other lakes are also affected by human activities, including diversion for agricultural purposes. The level of Walker Lake, for example, has dropped 132 feet since 1908.[6] And as Utahans know, wet periods can cause the Great Salt Lake to rise to levels that threaten roads and even the city's airport.

These fluctuations suggest a fundamental challenge that faces cartographers. They must depict constantly changing hydrological features like rivers and lakes as if they were fixed entities. True, one could draw a dotted line, rather than a solid one, around a lake to suggest that its perimeter fluctuates. More typically, however, cartographers draw a *solid* line instead to depict the feature as it appeared when experienced and/or mapped. This tendency to *delineate*—that is, literally draw solid lines around—something whose edges are not really stationary is characteristic of the cartography of western exploration.[7] Drawing these lines empowers not only the mapmaker, but also the sponsors and readers of the map. Lines delineate a feature for posterity by fixing its position. Moreover, by fixing its position we not only indicate it as if it were permanent; we also claim knowledge of the feature and thus position ourselves as claimants. This in turn may lead to a taking

FIGURE 1.3.

The Humboldt River near Palisade, Nevada. June 2001 photo by author

or control of that feature in both a scientific and a political sense. In actuality, one should regard lines on maps with considerable suspicion. Lines do not really exist as fixed in nature, yet cartographers, by their simple applications, portray a stationary world that awaits exploration and ultimate conquest.

Although of little or no value to navigation, the Great Basin's extensive but short-circuited river system is noteworthy. Whereas most of the region's streams are quite short, rising in the mountains and draining into the nearby adjacent valleys, there is one significant exception: The Humboldt River (fig. 1.3) rises in the mountains in the northeastern part of the Great Basin and flows westward for about 250 miles before disappearing into western Nevada's Humboldt Sink. The Humboldt's easy, water-level route made it a natural way westward in the nineteenth century, but it was also an enigma. Although it seems a fairly impressive river, especially in times of mountain snowmelt, the Humboldt carries insufficient water for it to cut through the formidable Sierra Nevada that form the Great Basin region's western boundary. Other perennially flowing rivers, like the Truckee and Walker, rise in the Sierra Nevada and flow eastward into the Great Basin. On the eastern side of the region, the Jordan River of Utah carries Wasatch Mountain waters northward from Utah Lake to the Great Salt Lake, where they slowly evaporate. Again, the region's aridity effectively interrupts, or rather terminates, the flow of water to the Pacific Ocean.

Paradoxically, however, it is water (in the form of either rivers or lakes) that creates many of this arid region's most prominent landmarks. Consider the irony: In an otherwise desolate area such as the western Salt Lake Valley, the intense blue waters of the Great Salt Lake define the landscape for about a thousand square miles. Yet the lake is legendary for its salinity. Its waters are so salty—several times saltier than the oceans—that bathers become almost supernaturally buoyant. Salt is commercially produced here—a reminder of the high evaporation rate in this desert region. Just when one is finally reconciled to the fact that waters in the Great Basin are generally unproductive, even devoid of life, another surprise appears: Surrounded by desert lands with scant vegetation, some of the region's lakes (e.g., Utah Lake and Pyramid Lake) contain fresh, life-sustaining waters teeming with trout and other freshwater fish. Then, too, even where water is not visible, its presence (or rather former presence) is suggested in the form of terraces or "benches" that mark the now-vanished shores of lakes that have receded since the Pleistocene period ended about 10,000 years ago.

Next consider the region's topography. Interspersed with salt-veneered or water-filled basins are spectacular ranges of mountains that tend to be oriented north-south throughout much of the region (fig. 1.4). These have been faulted and tipped downward as, over millions of years, the continent moved westward and split into fragments. Travelers and others have frequently commented on the region's mountains as they tend to be so regularly oriented roughly north-south, the exception being in the Mojave Desert, where mountains may be either isolated features (called "inselbergs," or island mountains), or even run transversely, that is, trend east-west. Like a Gordian knot in the center of North America's basin and range province, the Mojave Desert marks the southern end of the Great Basin on most geographers' maps.

Throughout much of the region today, vegetation is sparse, that condition also being a result of the low precipitation in this rain-shadowed area. Within the region, the topography in large measure determines the vegetation. Traveling through the Great Basin in 1859, Captain James H. Simpson was well aware that the better-watered mountains sustained trees. Western Nevada's Reese Valley, Simpson reported, was "exceedingly forbidding in

FIGURE 1.4.

Wheeler Peak and the Snake Range viewed from Spring Valley, Nevada.
April 2003 photo by author

appearance" and was in places "perfectly divested of vegetation."[8] The
mountains, however, presented a more cheerful prospect to Simpson. In the
same area, he noted that the "cãnons of this mountain abound in pure water
and splendid grass. The mountain mahogany is also seen." Simpson added
that "[c]edar and pines are also found, as they have been in nearly every
range since we left the Great Salt Lake Desert."[9] Most of the hills in this
region are devoid of trees, but where dependable fresh water flows, the ver-
dure can be stunning. "The contrast between the perfectly barren, sandy,
thirsty-looking country to be seen on every side and the valley of Walker's
River, fringed with green cottonwoods and willows," wrote Simpson, was
"very refreshing."[10]

Away from these river valleys, sagebrush *(Artemisia tridentata)* occupies
large areas. It is especially common in the northern and central Great Basin,
which is relatively high in elevation: the valley floors here are about a mile
above sea level. In the southern part of the region, the land is lower, valley
bottoms being about 2,000 feet above sea level west of sprawling Las Vegas.
That means that precipitation is lower and temperatures higher in the
southern part of the region than in the north. The region's ultimate
desert—Death Valley—is a trough whose bottom lies more than 250 feet

below sea level, the lowest and hottest place in the Western Hemisphere. In the southern part of the Great Basin, the characteristic olive-green creosote bush *(Larrea divaricata)* abounds. Throughout the Great Basin, then, vegetation is zoned more or less vertically. On slopes above about 6,000 or 7,000 feet, piñon pine *(Pinus monophylla)* flourishes. Above that zone, various taller pines, including ponderosa pines, and spruce grow. At the higher elevations, about 10,000 feet, the bristlecone pine makes a stand. Above this elevation, to the mountain heights of 13,000 feet as on Wheeler Peak, the land is bare—above the tree line—covered with shattered rock and snow.

The native peoples knew the land well, but it took outsiders generations, even centuries, to understand its varied character. Many of the Europeans who first saw this region immediately recognized its remote and forbidding quality but were so overwhelmed by its physical geography that they comprehended little else about it. To Europeans whose civilizations had developed around irrigated farming of crops, this was marginal land indeed. Its countenance was desertlike and its population sparse—just the opposite of the well-watered, populated areas that drew Spaniards in search of wealth and Indian labor. That forbidding and uninhabited quality helps explain why it took Europeans so long to penetrate, and ultimately map, the Great Basin.

# The Power of *Terra Incognita*
## 1540–1700

IN 1540, SEBASTIAN MÜNSTER stood back as he put the finishing touch-
es on a work of art—the beautiful map he titled *Tabula Novarum Insularum,
quas Diversis Respectibus Occidentalis & Indianas Vocant* (fig. 2.1). Working in
Amsterdam, Münster and his associates prepared this map using the latest
technology—a woodcut design that required them to carve the original
image in reverse. When ink was spread on the woodcut, and paper pressed
against it with considerable pressure, a crisp image resulted as if by magic
when the paper was lifted. On Münster's map, the western coast of North
America looks peculiar to us because we know it so well today from highly accurate maps and even satellite images. To us, Münster's coastline appears
misshapen and truncated, and we are surprised to see Zipangri (Japan) lying
less than two hundred miles off it. Then, too, the interior of the North
American West is extremely vague, with no recognizable features. Münster's
map appears primitive to us today. Back in Münster's time, however, this
map was state of the art and its accuracy seemingly above reproach. Münster
worked largely from narrative statements and other information from
explorers who had ventured to the margins of the known world. In creat-
ing his map, Münster yielded to a common understanding of his era—that
a large waterway cuts across much of North America. This waterway—
probably the mythical Straits of Anián—lured early explorers until, after
many attempts, they despaired of finding it. We strain to decipher the Great
Basin on this map and others of its time because that region simply did not
yet exist in the imagination.

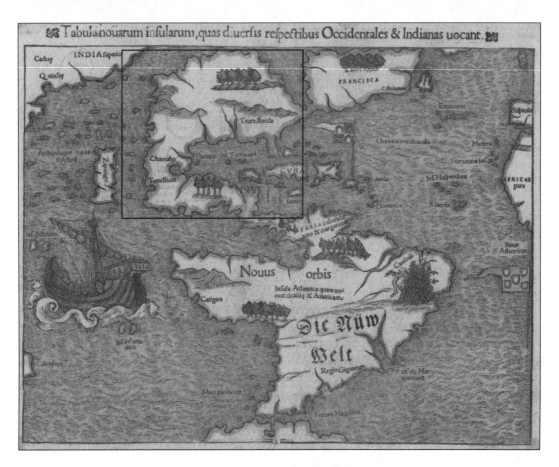

About twenty years later, when Girolamo Ruscelli completed his *Nueva Hispania Tabula Nova* (fig. 2.2), much had changed. Produced in Venice, Ruscelli's map was one of many that would shape Old World views of the New World. It seems more restrained than Münster's, and that is indeed the case. Among the earliest maps showing only a portion of North America, Ruscelli's is also more intricate than Münster's. It reveals the continent as a funnel-shaped outline into which rivers are incised and on which stylized mountains are heaped. Here and there, icons of villages indicate the locations of Native American and Spanish colonial communities. Ruscelli's map is both more cautious and less imaginative than Münster's. Its borders limit the amount of the continent shown, and this is not accidental. Apprehensive about the paucity of knowledge of what lay to the northwest, Ruscelli

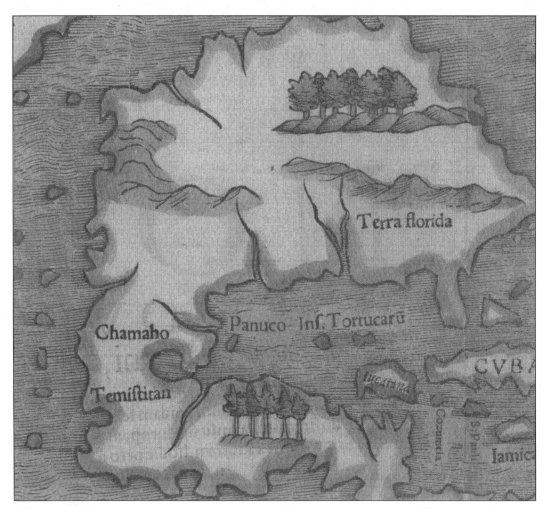

DETAIL OF FIGURE 2.1

excluded that area to the north that so deluded Münster. In the extreme northwestern corner of Ruscelli's map, the Rio Tontonteanc (today's Colorado River) drains a huge area completely devoid of features, natural or otherwise. This is the area we would much later call the Interior American West, and it remained a complete mystery to Ruscelli.

Like cartographers of all ages, Ruscelli used the maps of other cartographers for inspiration—in this case, Giacomo Castaldi's identically named *Nueva Hispania Tabula Nova* (Venice, 1548). At that time, such copying was not considered plagiarism, but rather an honor to the master whose work was copied. We can now see that Münster's map represented a more exuberant type of cartography than Castaldi's, but something else was also happening. At just this time, geographic knowledge was increasing so rapidly

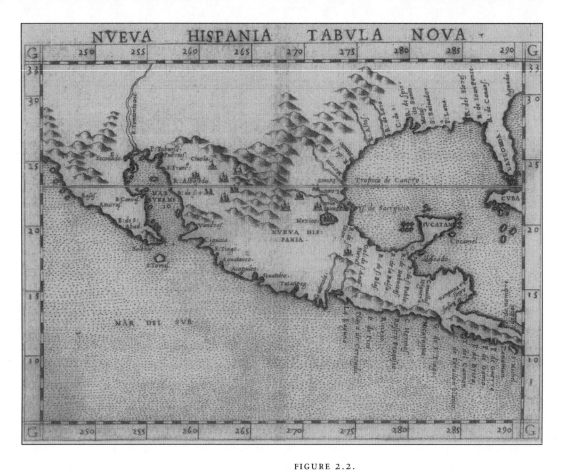

FIGURE 2.2.

Girolamo Ruscelli, *Nueva Hispania Tabula Nova* (ca. 1565).
Courtesy Special Collections Division,
University of Texas at Arlington Libraries

that it had rendered Münster's map obsolete. So it is that seemingly original maps may actually build upon some earlier maps and discard that which appears on others. These earlier maps provide not only a foundation, but also recognizable patterns of brick and mortar, metaphorically speaking, for the architecture of later maps.

Gazing at his entire map, Ruscelli must have felt a sense of satisfaction about how much of the area was known—mountains, communities, and rivers that he placed after consulting the records of explorers on bold *entradas* into the continent. As Ruscelli's eye roamed northward toward the top of the map, however, his satisfaction was likely tempered by the apparent incompleteness of his work. He knew something existed there but had no information about what it looked like and who—if anyone—lived there.

Like many cartographers of his era, Ruscelli could have festooned the empty spaces in this northwestern area with unverified geographic features or fanciful creatures and caricatures of exotic native peoples. But he held back, his restraint both satisfying and frustrating. If those empty spaces haunted Ruscelli, they did so partly because they indicated failures in exploration. And yet these same blank spaces were strangely satisfying because the unknown areas—*terrae incognitae*—meant that exciting challenges still lay ahead. The word *blank* is significant here because it is derived from the French *blanc,* "white." In a subliminal reference to virginity, a cartographic blank space equaled and translated into the purity of whiteness that was as yet unsullied by knowledge.

If Ruscelli was cognizant of the blank areas on the map, he was probably even more intrigued by what might exist *beyond* that northern border. Gazing there, he must have wondered how many natural marvels or wonderful settlements lay just beyond the tightly spaced twin lines that formed the rectangular border—the neatline—of his map. Would these areas long remain outside the known world, or would Ruscelli live to see them rendered in black and white, retrieved from their cartographic limbo and made *terrae cognitae?*

This question was answered within the decade, as the masterful *Americae sive Novi Orbis* (fig. 2.3), produced by Flemish cartographer Abraham Ortelius in 1570, depicted much of North America with growing accuracy, if not perfect fidelity. Based on information from European expeditions, Ortelius knew that the Pacific coastline stretched northwestward, even though he rendered it a bit enthusiastically as a huge bulge. Yet we sense that he helped configure the West in ways we recognize even today. Ortelius was not only a master mapmaker; he also produced the magnificent *Theatrum Orbis Terrarum,* which is widely recognized as the first atlas, or book of maps. *Americae sive Novi Orbis* confidently reveals the New World to an interested Europe. This map is, however, as stunning for what it does not show. In the interior of North America, it contains a huge blank space marked by the evocative words "Ulterius Septentrionem versus he regions incognite adhuc funt" ("Turning beyond [the] north, these regions are still unknown"). This was an honest enough assessment, at least from a European perspective. At this time, much of western North America lay beyond the margins of common knowledge. That area would only slowly reveal itself over several hun-

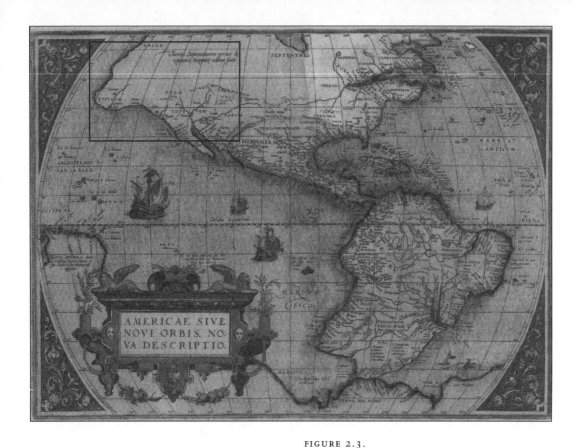

FIGURE 2.3.

Abraham Ortelius, *Americae sive Novi Orbis* (1570).
Courtesy Special Collections Division, University of Texas at Arlington Libraries

dred years, and only toward the end of that period would its mysterious heart—the Great Basin—finally be revealed.

Anyone who studies maps for very long arrives at an interesting but paradoxical realization, namely that they show both more and less than actually exists. Consider the case of the area that Münster, Ruscelli, and Ortelius were unable to yet comprehend—the interior of the American West. As historian David Weber notes of such challenges, "Attempting to understand a strange new world required mental adjustments that engaged and defied the best European minds."[1] As geographer John Logan Allen further notes, this was indeed the last major North American region to be fully explored,[2] but not simply by happenstance. This region was literally the farthest from the Europeans' actual experience and a blank on both their mental maps and their cartographic products.

But blank spaces excite the imagination and are too tempting to leave

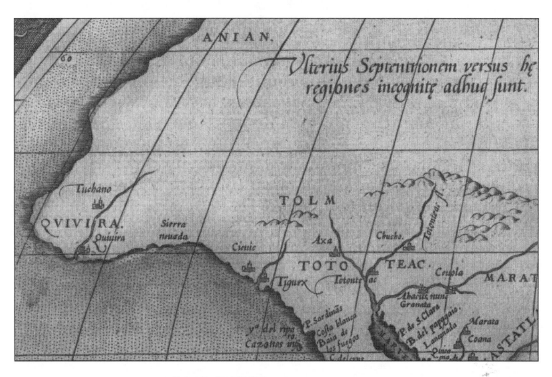

DETAIL OF FIGURE 2.3

blank for very long. By carefully studying historic maps of this region, we learn that they take shape slowly, often misrepresenting the actual geography in two ways: They may neglect to show geographic features that some explorers knew, or had heard on good authority, existed (for example, the Sierra Nevada), and/or they may depict features that do not exist at all, and never have in human history—like waterways flowing all the way to the Pacific Ocean from the interior of the continent. To decipher such maps requires not only a thorough knowledge of the actual geography, but also a recognition that belief is as important as observation in cartography. How else might we explain the persistence on maps of features that are widely understood to be inaccurate or even nonexistent?[3]

At this time, cartography was more art than science. Art is intuitively subjective, personal, and visual. Much as dreams inspire us in images rather than written words, maps began to shape the popular imagination visually. As a novelist put it, mapmakers at this time were "always dealing with secondary accounts, the tag ends of impressions." As a consequence, he concluded, "Theirs is an uncertain science."[4] That word *science* is crucial here, for it signals a sea change that was about to occur. Science conveys a new way of thinking: objective, verifiable by others, restrained. By the 1600s, with the development of ideas that would culminate in the Enlightenment, cartography begins its inexorable move from art toward science, with its emphasis on content rather than conjecture.

Both art and science suggest individual creativity and authorship, but a word of caution is in order. We commonly think that a map attributed to a particular mapmaker—Ruscelli, Ortelius, Mercator—was drawn or produced by that mapmaker. However, the production of a map is a complex process involving many people. Further considering this issue of authorship, what we call the "Ortelius map" is actually inspired by the maps of others. Ortelius was not only a master cartographer, he was also a prolific compiler of geographic information from diverse sources. Operating behind the scenes, or at least out of the limelight, Ortelius's collaborator Philip Galle was a force behind many of the varied products we attribute to Ortelius. We know little about some of the most influential cartographers in history, largely because their identities have been overshadowed by those of their sponsors.

Next consider the tenacious but human process of exploration, which inevitably affects the shape of maps. An enduring fact of continental exploration, its periphery-to-core orientation, helps explain why areas in the interior of continents remain *terrae incognitae* for so long. Inaccessibility results in one of the truisms about cartography in the Age of Exploration: Generally, the coasts were fairly well understood and depicted, but accuracy of cartographic information diminished rapidly inland. The reason for this discrepancy is a function of the methods used by explorers to gain access to places: As cartographers crafted maps of North America over several centuries, the continent's general outline quickly took shape first, followed by a much slower articulation of interior spaces. In this regard, North America was much like other continents, including South America or Africa. There, too, Europeans using water-based vessels of discovery first comprehended the continents' peripheries, leaving the task of entering interior spaces to later explorers. These later expeditions, or *entradas,* required either watercraft able to ascend rapidly flowing rivers or foot-based expeditions capable of traversing great distances. Five hundred years later, Umberto Eco would summarize the dilemma facing mapmakers in the 1500s: "He found it: large, colored and incomplete, as many maps were then incomplete out of necessity; the navigator, coming upon a new land, drew the shore he could see but left the rest unfinished, never knowing how and how much and whither that land extended."[5]

Overall, North America, like other continents, generally came to be understood in stages; first, by her oceanic margins, next by her riverine ribbons of navigable water, and finally by overland expeditions into her landlocked heart. The glaring exception appears to be accidental indeed—the shipwrecked Alvar Núñez Cabeza de Vaca, who in the early 1530s trekked overland in the company of three compatriots. Traveling generally westward and southwestward through southwestern North America in hopes of finding Mexico City, Cabeza de Vaca left the first written account of interior exploration, albeit considerably south of the Great Basin.[6]

Water buoyed and nurtured explorers and also had a role in defining both exterior form and interior design. This role is evident in early maps of North America like Münster's, many of which embody the cartographic consequences of the search for the enduring Straits of Anián and the Northwest Passage. A historian of exploration calls these searches for fabled waterways "voyages of delusion."[7] They were motivated by a combination of factors—the prospect of handsome rewards for those who first located a waterway through the continent, persistent myths based on folklore rather than experience, and outright wishful thinking. Water passages through the Interior West are common on some Spanish maps of this period, but they are more common on British and French maps. Given their penchant for waterborne travel and commerce, it is not surprising that the French succumbed to tales of water routes across the entire continent. On some French maps, the Rivière Longue (Long River) is prominent, and on others a huge interior sea called Mer de L'Ouest (Sea of the West) occupies much of the area where the Great Basin is located. No less a rational cartographer than Guillame de l'Isle is reported to have depicted this mythical feature. However, one French cartographer in particular—Baron Louis Armand de Lom d'Arce Lahontan—immortalized these fanciful western water bodies on his eighteenth-century maps.[8] The belief in a water body through the continent was remarkably long lived. On numerous maps, and in popular thought, it persisted for nearly three centuries. Although quite late in the Age of Exploration, Lewis and Clark's primary emphasis on aquatic transport typifies the enduring search for (and travel on) waterways. That Lewis and

Clark's Corps of Discovery searched for a water passage through the continent in the early nineteenth century reveals how long myths can persist.

This fruitless three-century search is comprehensible only if we understand, European aspirations in light of the environment. In reality, Interior Western North America presented a set of unique conditions that required overland trekking where waterways were scarce, absent, or led nowhere. Although exotic streams like the Rio Grande and the Colorado River drain portions of the Interior West to the south, and the Columbia River drains a huge area to the north, much of this vast interior is simply not reachable by water at all. If much of the Intermountain West is a land of little rainfall and stark landscapes, its deepest interior is the driest and most landlocked of all. It is this portion of western North America that both eluded and attracted the explorer. As if to confirm the Great Basin's inherent isolation to an inquiring outside world, it is worth noting that there is absolutely no record of any European setting foot in it for more than two hundred years *after* Cabeza de Vaca's journey in the 1530s and Francisco Vázquez de Coronado's expedition in the 1540s.

In exploring western North America, Spanish *entradas* of the 1540s faced several challenges. The first was psychological, for they had inherited a medieval mindset involving considerable romance about fabled lands that always seemed to lie just out of reach. Shortly after conquering Mexico in 1519, the Spaniards moved northward into the area of wild tribes and desolate landscapes they called "Chichimeca." Their well-armed expeditions into the north soon enlisted the support of native peoples, who supplied additional arms and geographic information. These Spanish *entradas* hoped to locate an "Other Mexico"—a vision manifested in the belief that the Seven Cities of Cíbola in the kingdom of Quivira existed somewhere in western North America.[9]

In addition to their willingness to believe fanciful stories about fabulous cities and lost empires waiting to be discovered (or rather rediscovered), the Spanish explorers also faced a dilemma that confounded all explorers of this age: Unable to accurately determine *longitude* (the distance east or west of a particular point), they created maps that involved considerable guesswork. Whereas determining latitude (the distance north or south of a particular point) is relatively easy if one knows the day of year that the measurement is made on, and hence the position of the sun at high noon and therefore one's distance from the equator, longitude requires knowing the exact time of day in two places simultaneously. One needs to know the distance between the place where the explorer is and the place or point from which all measurements are made, which might be Seville or London. That, however, can be determined only by knowing the exact time in that far-off location. Calculating longitude became possible only after the invention of the chronometer, and that invention lay two centuries in the future.[10] In the meantime, the Spanish explorers did their best to collect geographic information via fragmentary field maps that found their way back to Spain, where they were in turn added to a master map that was guarded with utmost secrecy by the Casa de Contratación in Seville.

Despite, perhaps because of, such obstacles, the Interior West possessed considerable allure. Consider this region's growing appeal that was based on mystery—the seductive *terra incognita*. How does one depict that which he has never seen but someday hopes to know? In both words and images, the depiction of blank spaces has long enchanted students of cartographic history. These are more than blank spaces: they are essentially "silences," in that language stops here. Just as we strain to hear what lies within those silences, we long to see that which is hidden from view. The imagination can work overtime in such spaces. Exploration and discovery are juvenile in that they are perennially associated with youth. Metaphorically speaking, they are associated with the exuberance of a culture's youthful discoveries that in turn led to the more serious side of colonization. Recall the words of Marlow in Joseph Conrad's *Heart of Darkness*: "Now when I was a little chap I had a passion for maps. . . . At that time there were many blank spaces on the earth, and when I saw one that looked particularly inviting on a map (but they all look that) I would put my finger on it and say, When I grow up I will go there."[11] Through the process of exploring those blank spaces, knowledge grows as it destroys the naiveté of youth. That process would ultimately take more than two hundred years, however, and it is to that drama we return.

Continuing our story from the time of Ruscelli and Ortelius, let us observe how cartographers depicted the unknown in the century *before* the first Europeans, which is to say Spaniards, set foot in it. Given the Europeans' penchant for water-based exploration, the relative lateness of this landlocked area's discovery and exploration is not surprising. Because mapping tended to follow exploration—that is, to work its way inward from the coastlines of discovery—the first maps showing the interior do so only by implication. These maps confirm that the area was peripheral to Spain's interests at an early date. To early explorers, the area was certainly mysterious and seductive while simultaneously frightening. Because Cabeza de Vaca's 1530s exploration was both accidental and unmapped, it was incumbent on the first deliberate *entradas* of the 1540s to confirm some of the tantalizing information from the earliest trek through the region. Even Coronado's ambitious *entrada* of the early 1540s, however, was deflected to the northeast by a geographic obstacle (the formidable Grand Canyon of the Colorado River) and cultural misinformation (the lure of cities of gold elsewhere as described by Indian informants all too anxious to get the Spaniards to move on). Early expeditions therefore failed to enter the area we call the Great Basin. Coronado and his men seemed humbled by the rugged arid country to their northwest just as they were beguiled by tales of wealth to the northeast. They cursed the Grand Canyon, which stopped them in their tracks; but because they sought riches that allegedly existed in another direction, that deflection had far-reaching consequences. It left the Great Basin untouched.

Throughout all of the seventeenth century and part of the eighteenth, the interior of western North America remained perfect virginal *terra incognita*. This unknown area soon took on another quality, for *terra incognita* was often *tabula rasa*—a blank slate ready to be embellished. Lack of actual information about the region did not stop some from depicting the locations of rivers far inland, even though explorers may not have personally seen these features. In reality, the western part of North America was quite fluid, cartographically speaking. Small wonder that some maps also depict fanciful hydrographic features—fanciful at least to our eyes today. These features more or less corresponded to what one might call topomyths: that is, features that were widely *thought* to exist in the absence of proof to the contrary.

At this time, many Europeans refused to give up on the vision that a water passage existed to the Pacific at the western edge of North America; thus, it is not a coincidence that many maps of the region at this time still exhibit a transversely positioned body of water that appears to be either a Northwest Passage or, as the Spaniards once called it, the Straits of Anián. Significantly, this mythical waterway lies well north of the Great Basin, at about 49° north latitude. Its closest counterpart in reality may be either the

Strait of Juan de Fuca or the Columbia River. Generally, however, much of the Interior West remained a blank space that increasingly irritated those aiming to reach the Orient by traversing the North American continent. It was this vision of transoceanic trade, rather than development of the continental interior itself, that motivated the next generation of explorers and mapmakers. That impulse seemed to seal the cartographic fate of a land with no rivers that reach the sea, for the very idea that there could be such an area defied both logic and imagination.

It is here that we must take a closer look at what appears, and what doesn't, on those maps from the period 1540-1700 that include the Great Basin. The region may take various forms, but one should be above all impressed with the concept of void or emptiness. It is either shown as a complete blank, or obscured in some way, or marked with the mysterious words *terra incognita*. Let us, therefore, next consider *terra incognita* not in space, but in time. It occupies a crucial niche chronologically as well as psychologically. It is tempting to think of *terra incognita* as an eternal concept that is rendered obsolete by the Age of Exploration. After all, the Latin roots of this term suggest ancient knowledge. But in fact, *terra incognita* is a concept of the Age of Exploration itself, and that age only began in the mid-to-late 1400s. In maps of the medieval period, remote locations were often festooned with fanciful creatures and exotic peoples with supernatural features. The Age of Exploration witnessed a decrease in such oddities as explorers began to report phenomena more and more accurately. The use of Latin as a universal language for knowledge was endorsed at the end of the Middle Ages, and so the term *terra incognita* perfectly captured both the spirit of the times and the character of unknown places.

The widespread use of Latin coincides with another development that would revolutionize cartography, notably the resurrection of maps described by Claudius Ptolemy, a Greek philosopher and geographer from Alexandria, Egypt. This rediscovery of Greek cartography occurred at just the time that explorers began to aggressively penetrate new lands. However, it was not based on an actual map, but rather on Ptolemy's narrative *about* a map; this sufficed in an age when classical authorities were revered. In the 1500s, European exploration required a system that could bring order to the varied discoveries that were under way. Maps were not coincidental, but essential, to this process. The new age of exploration called for new maps.

In contrast to most medieval *mappae mundi*—which featured the holy city of Jerusalem at the top (which is to say their orientation was *east*) and divided the known world into three portions (the Near and Far East at the top; Africa to the right; Europe to the left)—maps by Ptolemy were oriented with north at the top and land masses more or less zoned latitudinally. This gave rise to the convention we now take for granted, the North Pole as up, the tropics as below, and the South Pole as bottom. It also gave rise to the midlatitudes between the frigid Arctic and torrid tropics. It is in those midlatitudes that the region we call the Great Basin lies. Like many places in these latitudes, especially those in elevated continental interiors, the Great Basin is subject to huge seasonal temperature swings, windy spells, and extended periods of drought.

I have suggested that the new Age of Exploration called for new maps, but that suggestion needs to be qualified. Maps represent not only graphic images, but also words that work in concert with those images. Although no maps by Ptolemy existed in the Age of Exploration because the great fire at Alexandria had destroyed them in A.D. 391, Ptolemy's *words* helped scholars in the 1500s reconstruct, and hence perpetuate, his lost map. The fact that the original was lost also enhanced its intrigue to an age searching for an elusive past. Ptolemy's Greek origins sat well with an increasingly learned culture that looked to the ancients for wisdom as the Middle Ages drew to a close. Latin, which was kept alive by the Church, soon formed the lingua franca of the educated. Thus it was that Ptolemy's map suggested tradition and accuracy while at the same time offering new prospects for cartographers in the Age of Exploration. Every map illustrated in this book owes a debt to Ptolemy, or rather to the rediscoverers of the writings of Ptolemy.

Ptolemy's geography possessed yet another advantage that suited the Age of Exploration. With its general zonation to suggest lands of varied climatic character, Ptolemy's map superimposed a reticulated grid—a series of evenly spaced lines crossing at right angles—the forerunners of latitude and longitude. This reticulated grid was revolutionary, for it enabled the location of specific places anywhere on earth to be determined with accuracy. Those points could be determined by using the position of the sun, which in turn determined the seasons. Ptolemy's map ultimately assumes knowledge of basic astronomy and is in turn based on an awareness that the earth itself is a sphere. Popular suppositions to the contrary, this was common knowledge in Columbus's time, especially among learned people. By the Age of Exploration, the world was known to be round, and globes were commonly used as both objects of curiosity and devices of conquest.[12]

This period witnessed the rise of scientific mapmaking with its demand that phenomena be verifiable (and verified) and accurately located. Maps, and books of maps (atlases), became more common after around 1580. Still, certain challenges remained. The age-old dilemma of representing a spherical surface—the earth, or curving portions of it—on a flat piece of parchment still faced mapmakers. The main challenge, how to account for the distortion that inevitably follows opening up a sphere and laying it out flat, remained. Resolving this issue resulted in a wide array of map projections, the most popular and enduring of which was that by Gerardus Mercator in 1569. Mercator was a friend and associate of some of history's greatest mapmakers, including Abraham Ortelius. Because Mercator's projection imposes a grid from the equator to the poles, it notoriously exaggerates the size of polar regions. However, because it maintains fidelity to the grid, all features depicted are true to each other as regards direction. By the late 1500s, the combined influences of Ptolemy, Mercator, and others dominated mapmaking: most maps of exploration were now oriented with north up, and most employed a grid system that enabled users to locate places with accuracy by pinpointing their positions with regard to latitude and longitude. This was the state of the art as maps showing Interior Western North America proliferated, even though the shape of that region continually morphed as new information came to the attention of cartographers.

If these maps of early Renaissance Europe seem artistic, there is good reason. In the 1500s, cartography and art, especially painting, began to flourish at about the same time. Maps are not only seen in paintings of the period as props and symbols, but the artistic conventions themselves show up in maps as color and composition become important elements. Cartographic historian David Buisseret observes that there was a "smallness and intimacy" among learned people, and that painters and cartographers had "frequent informal contacts, which must have encouraged interchange between 'artists' and 'mapmakers,' blurring the line between them."[13] These maps appear as much works of art as of science, and they indeed are. Like all maps, their design reflects not only technology but also the sensibilities of the time.

Consider next the case of the Franciscan friar Vincenzo Coronelli's

*America Settentrionale* (fig. 2.4), which was produced in Paris in 1688. As cartographer to the Court of Venice and the French king, Coronelli had a wealth of sources upon which to base his work. French sources were especially influential, but Spanish information, or rather misinformation, was also used. Although published more than a century *after* maps by Ruscelli and Ortelius, Coronelli's map boldly depicts California as an island, a reminder that neither geographic knowledge nor maps develop in a perfectly progressive manner. This interesting exception to the rule that the periphery is known first (and best), reminds us of two things: 1) how peripheral that part of North America still was in the affairs of most Europeans in the sixteenth and seventeenth centuries, and 2) how difficult it was to get deep enough inland to confirm that California was part of the mainland.[14] Some historians suggest that Spaniards may have deliberately provided disinformation as a way of confusing those who quested and lusted after Spain's New World possessions.[15] Of special interest to us is the area of the Great Basin, which Coronelli covers with a large block of text that reveals its ongoing nature as, he himself calls it, "terra incognita." That area lay north of the best known part of Spain's frontier, New Mexico. However, reading beyond Coronelli's words, or rather between Coronelli's lines, we note that he has nevertheless stamped it with European words—a precursor to claiming the area at some future date.

The western part of Coronelli's map was based on information from Diego Penalosa and Nicolas Sanson d'Abbeville. Penalosa was governor and captain-general of New Spain, and Sanson was known as "the father of French cartography"[16]—a vivid example of the international influences on seventeenth-century maps. Despite this collaboration, authority, and talent, much of the mapped area is nevertheless more conjectural than conclusive. And yet, at just this time, natural science began to flourish, and with it demands that observers record only what could be observed—and verified. David Buisseret notes that "the seventeenth century was also the time when natural scientists naturally turned toward maps . . . to explain their theories." As this contact intensified, "the natural sciences came to rely on maps for the assessment of spatially based evidence." This period has been described as "the age of scientific cartography," but it especially involved a refinement of cartographic methods, which in turn depended on improvements in technology.[17] In reality, unless an expedition had scientists traveling with it, consider-

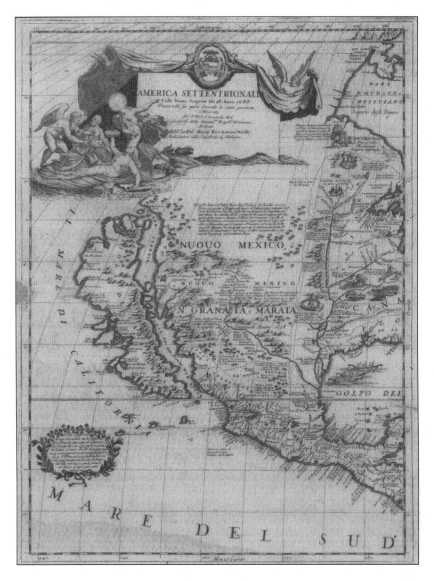

FIGURE 2.4.

Vincenzo Coronelli, *America Settentrionale* (1688).
Courtesy Special Collections Division, University of Texas at Arlington Libraries

able conjectural information was liable to appear on maps resulting from it. In the Interior West, the coming of science was considerably delayed. This was a marginal area, and Spain applied her best talents elsewhere.

Maps are never separable from politics, and that fact deserves closer inspection at this time. Spain was the major political force here, and so one would naturally expect most maps of the 1500s and 1600s to be Spanish. Not so. In point of fact, most of the printed maps showing this part of

North America were produced in Venice, Paris, Amsterdam, even London. Spain may have had the greatest knowledge, for she launched numerous expeditions to western North America in the late 1500s and 1600s. However, because of concerns about interlopers and marauders like the ambitious English pirate cum naval officer Sir Francis Drake, Spain kept most of her geographic secrets close to the vest. Spain was a notoriously sparse producer of maps while, ironically and irritatingly, the maps of her possessions done by Italian, French, Dutch, and English cartographers abounded. Drake's legacy is evident on the Henricus Hondius maps of the New World prepared in the Netherlands. On these maps, California bears the name bestowed upon it for Queen Elizabeth by Drake: New Albion. Significantly, the interior of New Albion is not articulated, as it lay beyond the experience of English explorers. For her part, Spain produced virtually no maps for public consumption. Those maps of New Spain were most often prepared by others, printed—that is, produced in large quantities as essentially identical copies—and marketed to an eager public. At this time, Spain's maps remained largely as manuscripts—hand-drawn, one-of-a-kind originals that were kept under lock and key in archives and military drawing rooms.

Thus it was that the Age of Exploration produced maps of two kinds. One type was reproduced widely and shaped the popular geographic imagination. The other, seen by a privileged few, was often more accurate. Unbeknownst to the public, these manuscript maps were used to fill in the blanks on a master map, called a "mother map" by some scholars. This master map contained the most accurate and up-to-date, which is to say sensitive, geographic information. Still, by around 1700, the area that would become the Great Basin was rendered in only the sketchiest detail. Overall, maps until about this time reveal that the region was so little known that it provided the perfect place to put elaborate cartouches. Often brimming with ornate detail and orienting the user to the map's scale and its legend, these cartouches conveniently hid the most basic reality of all: that no European knew anything about a huge, and possibly crucial, piece of northwestern New Spain. With the passage of another few decades after 1700, that situation would begin to change quickly and radically.

# 3

## Maps and Early Spanish Exploration
### 1700–1795

MAPPING, LIKE EXPLORATION, does not proceed evenly over large areas. Although portions of what we would later call the Southwest, especially New Mexico, were rapidly explored, mapped, and even settled in the 1600s, the Great Basin was not. That discrepancy is attributable, for the most part, to two factors: the Rio Grande Valley, which served as a corridor for Spanish infiltration, and the distribution of Indian peoples, who attracted the Spaniards like a magnet. In the Great Basin, native population density was light and total numbers small, especially compared with two rather densely populated (and hence attractive) areas—1) the irrigated pueblos along the upper Rio Grande near Santa Fe, and 2) fish-rich and oak/acorn-endowed coastal California. As noted earlier, the native peoples in the Great Basin were either nomadic or lived in very small, seasonally occupied villages. As a consequence, much of their homeland remained out of the Spaniards' reach. The stupendous physical geography bordering much of the Great Basin, with its deep canyons, tall mountains, and searing deserts, also helped keep this region a secret. Encountered only by a grueling overland trek through desert from the south or by crossing formidable mountains on its eastern or western margins, the Great Basin remained no-man's-land, at least on paper.

However, things were about to change. By the early 1700s, even before the Spaniards had actually explored the Great Basin, an important feature begins to appear on Spanish maps of western North America. Significant-

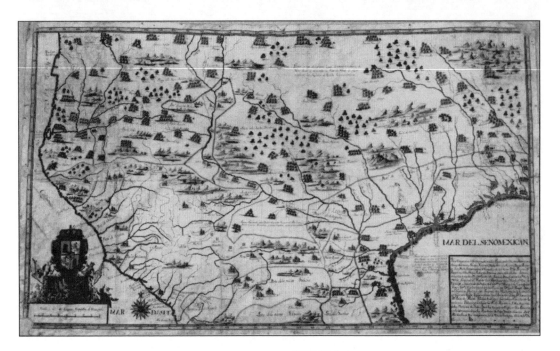

FIGURE 3.1.

Francisco Barreiro, *Plano Corográphico e Hydrográphico de las Provincias . . . de la Nueva España* (1728). Courtesy Hispanic Society of America, New York

ly, it bears an Indian name—*Timpanogos* or some derivation thereof—suggesting that Indian information about it preceded the Spaniards' actual arrival. A large unnamed interior lake first appears on manuscript maps by the early eighteenth century, notably Francisco Barreiro's beautiful *Plano Corográphico e Hydrográphico de las Provincias . . . de la Nueva España* (fig. 3.1). The words *chorographic* and *hydrographic* in the map's title are significant. They suggest a developing scientific interest in the surface of the land and the configuration of its waterways. On this intriguing map, produced in 1728, streams seem to flow *toward* an interior lake, suggesting that this is a region of interior drainage. On other maps of this same time period, however, the rivers appear to flow *away* from this lake, which has several names (including Youla and Timpanogos). This ambiguity is revealing, for it suggests just how little was actually known about the hydrology of the Interior West. Several other lakes in the region, including Bear Lake and Pyramid Lake, may have served as the prototype for this striking cartographic feature. However, Utah lake is most likely Youla Lake, although the Great Salt Lake is also a strong candidate because it is the largest body of water in the Interior West.

To those familiar with Spanish *entradas,* the appearance of any hydrologic features on these early-eighteenth-century Spanish maps is perplexing. There is no record of actual Spanish exploration into the Great Basin at this time, although *entradas* had reached the southern San Joaquin Valley of California—a semiarid, mountain-surrounded, low-lying area of interior drainage that gave them a foretaste of what would be encountered farther inland. There is, however, a distinct possibility that the Spaniards learned about the Great Basin from the natives in coastal California who traditionally traded with Indians living to the east of the Sierra Nevada.

But what about those native inhabitants of the Great Basin? Did they map the region, or parts of it? It is here that we must further broaden our definition of a map. Until recently, it was thought that the native peoples here had no maps, and some even suggested that native peoples were too primitive to create maps. Although it is true that Native Americans did not have the kind of maps with which explorers were familiar—increasingly scientific charts on paper that depicted places using graticules and latitude and longitude—that does not mean they were unaware or incapable of mapping. Consider, instead, that Native American mapping was far more intuitive. The natives had little use of mapmaking in the form of charts, and the landscape was more likely integrated into a cognitive mental map. As a cartographic historian observed, "for many American Indian peoples, the land was often its own best map—almost as if the topography itself possessed some sort of volitional authority."[1] That volitional authority was likely linked to Native Americans' strong belief in the inherent power of places to convey the power of the Great Spirit, but it could also function in more mundane ways. Each particular place was associated with storytelling; the interconnection of these places by memory formed a fascinating juncture between what we widely separate in Western thought—history and geography.

Historian of exploration D. Graham Burnett makes a very important observation about explorers' relationships to native peoples: Entering native lands meant encountering not only native inhabitants, but also their spirits. Burnett observes that "the encounter with Amerindian spirit landscape was an inescapable part of any penetration into the interior, and the dynamics of that encounter were essential to the process of exploration." He adds that "the terra incognita of the interior, though sparsely populated with Amerindians, was seen by Europeans as thickly colonized by Amerindian spirits."[2] So it was in the interior American West, where the explorers' reports often make mention of these native "superstitions" that affected everything from morale to logistics in the process of exploration and discovery.

The natives' attitudes ultimately affected explorers' mobility, for even the most intrepid of explorers needed indigenous peoples to guide them as they moved into new territory—both physical and cultural. As Burnett correctly observes, "Where Amerindians would not go, European explorers could rarely penetrate."[3] This is a theme that appears repeatedly in the Great Basin's cartographic and scientific history. Explorers would feel confident enough to go it alone only when they had initially mastered the lay of the land, and that recognizance, in turn, occurred with the help of native peoples. When hearing of geographic features, the Spaniards often used Indian names for them; in other words, the landmarks of Indians became the landmarks of the Spaniards through a subtle form of appropriation. Because so much rested on the explorers' authority, however, native peoples were rarely credited with providing geographic knowledge. To do so would have undercut the explorers' power.

Sometimes, however, the evidence of this exchange is hidden in the maps themselves. Tellingly, Barreiro's 1728 map suggests not only considerable native input of information, but also adoption of actual Indian cartographic technique, as in the "tadpole" lake configuration that cartographic historians believe to be "a strong indication of Amerindian source."[4] Here the Spaniards had the advantage. Although "Native Peoples throughout North America made maps indigenously, certainly from early contact times and probably long before that,"[5] their maps were vastly different from those created by Europeans. Whereas mapmaking may be a universal trait shared by all peoples, we should not expect their maps to look anything like one another's. Native peoples in the Great Basin had no reason to draw maps of the entire region, for only the Europeans' sense of empire called for that kind of elaborate, geopolitically inspired project. Cartographic historians David Woodward and Malcolm Lewis observe that "formal mapmaking (the inscription of spatial knowledge) tends to arise as a discourse function only within highly organized, bureaucratic societies."[6] Often organized around small bands of a few dozen people, Native Americans in this region had no such elaborate societal structure. They certainly had no scientific inclination to render maps accurately as regards directions from point to point or the shapes of geographic features. That was a European obsession, and it affected maps done in the 1700s, as this was the Age of Enlightenment dominated by reason.

Consider the dilemma that Spain increasingly faced on its North American frontier. She wished to encourage colonization but was apprehensive about how much information should be revealed. Overall, the region's low population density was more of a curse than a blessing to the Spaniards. Whereas the densely settled empires of Mexico and Central and South America were relatively easy to conquer, the lightly populated areas were not. Their inhabitants tended to be nomadic, free-spirited, and notoriously hard to control. Consider, too, Spain's dilemma geopolitically vis-à-vis other ambitious European powers. Cartographic historian Dennis Reinhartz notes that several factors, including both the sparsity of Spain's settlements and her reluctance to make public claims via maps, "encouraged Spain's European rivals, France and Great Britain, to engage in real and cartographic 'filibustering' campaigns in the region in the eighteenth century."[7]

There were some exceptions to Spain's intense cartographic secrecy, but they were few and far between. By the mid-1700s, a few Spanish cartographers published maps and atlases. These included Tomás López de Vargas Machuca, who in 1760 established the only independent cartographic publishing house in Spain. In collaboration with Juan de la Cruz Cano y Olmedilla, López produced several maps of portions of New Spain's northern frontier. López became "geographer to the king" in 1780 and was even authorized to create a geographic agency for the secretary of state in 1795. Both López and Cano had studied in Paris under one of France's premier cartographers, Jean Baptiste Bourguignon d'Anville. Not surprisingly, their maps reflect his influence. And yet their maps also "reflect the restraints of state secrecy." Reinhartz concludes that "in the middle of the eighteenth century, even the Royal Geographer was allowed to go public with only so much Spanish geographic information."[8]

Throughout the 1700s, the Great Basin region began to be better understood by European American explorers. As might be expected, the Spaniards first approached and entered it from the south or southeast, notably from Santa Fe, New Mexico, which was well settled by this time and served as a major outpost on Spain's northern frontier. Searching for a land route to the Pacific—more specifically from Santa Fe to Monterey, California—the Domínguez-Escalante expedition represented Spain's first serious attempt to traverse the region. Faced with the daunting deserts to the west, the expedition was thwarted in its attempt, leaving a candid assessment of the inhospitable environment there. And yet, despite the expedition's seeming failure, these Spaniards drew maps of portions of the region that reveal a struggle to reconcile firsthand observation with the stories provided by native informants.

During the late 1700s Spain's cartographic knowledge made great strides, especially compared with other European maps that seem oblivious to Spain's growing awareness of the interior. For example, on the majestic *Carte de l'Amérique Septentrionale* by Jacques Nicolas Bellin (Paris, 1755), the Interior West shows as largely blank space. The words appearing here, "On peut placer ici les Provinces de QUIVIRA et TEGOUAIO dont on na aucunes connoissances certaines" (One can place here the provinces of QUIVIRA and TEGOUAIO, about which one does not have any certain knowledge), reaffirm the long-held belief that an elusive empire of native peoples existed somewhere in the interior.[9] That contrasts with Spanish maps showing a region festooned with enigmatic lakes and rivers that seemingly ran backward.

Spaniards' dreams of empire in the interior yielded to reality as they aggressively explored their northern frontier in the 1770s. The map produced by the Domínguez-Escalante expedition was one of the landmarks in the cartographic history of the North American West. It would influence other maps for more than a generation. It focused attention on a crucial element—hydrography—that would be essential in helping later mapmakers depict the region. As the expedition explored near Utah Lake and the Great Salt Lake, its members were heartened by the presence of water. The well-watered, marshy valley, in fact, was said to compare favorably with the Valley of Mexico—an analogy seemingly aimed at interesting Spanish leaders because it suggested large numbers of native peoples. According to the expedition's leader, Captain Bernardo de Miera y Pacheco, this location was "capable of maintaining a settlement with as many people as Mexico City, and of affording its inhabitants many conveniences, for it has everything necessary for the support of human life." Note that this description represents a departure from earlier Spanish beliefs that fully developed empires existed in this area; instead, it suggest the *future* potential of the locale. Today the sprawling Salt Lake City metropolitan area stretching north and south along the Wasatch front seems inevitable, but Miera's prophecy deserves credit as visionary; however, typifying Spain's ambivalence and general mis-management of the northern frontier, officials failed to follow up on it.

Miera knew that abundant fresh water was a key element in sustaining the population here, and to him it was a perfect oasis. Utah Lake, which Miera overestimated to be six leagues wide and fifteen leagues in length, was a crucial factor in influencing the expedition's enthusiastic report, which noted: "This lake and the rivers that flow into it abound in many varieties of savory fish." By contrast, Miera gave a less sanguine report on the Great Salt Lake, observing that "[t]he other lake with which this one communicates, according to what they told us, covers many leagues, and its waters are noxious and extremely salty, for the Timpanois assure us that a person who moistens any part of his body with the water of the lake immediately feels an itching in the part that is wet."[10] Significantly, Utah Lake and Salt Lake are incorrectly depicted as one large, interconnected, hourglass-shaped feature called "Laguna de los Timpanogos" on the map, *Plano Geographico de la tierra descubierta, nuevamente, a los Rumbos norte, noroeste, y oeste del Nuevo Mexico* (fig. 3.2), that was inspired by Miera. On Miera's map the Rio de S. Buenaventura flows from northeast to southwest into Laguna de Miera. This, too, is incorrect. Miera's Rio de S. Buenaventura is likely the Green River (which flows into the Colorado River, not the Great Basin). Where Miera's river reaches the lake that bears his name, he likely depicted the Sevier River, which flows into Sevier Lake. Equally significant is another hydrographic feature on this map: In its upper left-hand corner, the Rio del Tizón flows westward from the Great Salt Lake, presumably all the way to coastal California[11]—at least that is exactly what Miera speculated in his report to the king in 1777.

Most significant of all, perhaps, is the fact that the Great Basin itself occurs at the far northwest corner of Miera's map, and a portion of it is can-didly labeled "Tierra Incognita." In his 1777 report to the king of Spain, Miera's description of this region's waterways is noteworthy. The report downplays, actually omits, reference to the adjoining noxious salt lake, while retaining the glowing language about the freshwater lake's resources. Miera also inflates in importance the region's rivers, noting that the "river which the inhabitants say flows from the lake, and whose current runs toward the west, they say is very large and navigable." Miera conjectured that the Rio del Tizón might be the river "discovered long ago by Don Juan de Oñate,"

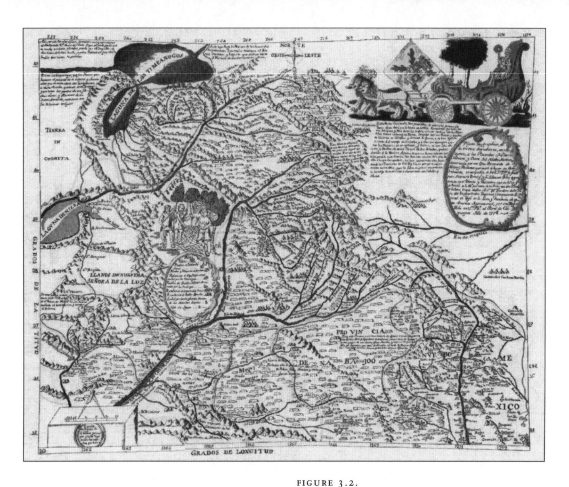

FIGURE 3.2.

Bernardo de Miera y Pacheco, *Plano Geographico, de la tierra
descubierta . . . del Nuevo Mexico* (1776).
Courtesy Utah State Historical Society

which that explorer "was not able to cross . . . on account of its great width
and depth." Miera recommended that the lake and river be explored, for by
so doing "in a short time a very beautiful province would be formed, and it
would serve to promote and supply the nearest ports of the coast of
California."[12]

The names that Miera gave to features are noteworthy. Some were Indian names, but others were Spanish. Consider, for example, the Rio de S.
Buenaventura—a name in which the melodious *buenaventura* literally means
good luck or good fortune but also suggests ease of travel. The name
Buenaventura was prophetic, for it became the icon of future explorers who
believed that a river might someday convey fortunate travelers westward to

the Pacific. Miera's map that accompanied his terse but informative report illustrates a portion of this river, and on it he again notes that this may be the river that Juan de Oñate encountered but was not able to cross, due to its width and depth; this is yet another suggestion of a river's potential importance in a land so otherwise devoid of water. A trail (often called the Old Spanish Trail on later maps) would ultimately lead from Santa Fe across the southern margins of the region near present-day Las Vegas; that trail reached the coast near Los Angeles. But the dream of the Domínguez-Escalante expedition to cross the region into northern California would remain unrealized—thwarted by both the intimidating physical environment and the immense logistical and political difficulties on Spain's far northwestern frontier.

Miera's map offers some important lessons in both history and historiography. To better understand this, consider not only its effect on maps of the same time period, but copies of the map itself. I have discussed Miera's map as if it were one original, but there are actually *seven* recognized manuscript versions of this map in various collections.[13] In geographic outline and content, all of these versions appear similar at first glance. Miera's original is likely in the British Museum, which cartographic historian Carl Wheat notes is understandable given the British officials' penchant for collecting maps and charts of all types, as well as the value of such maps in military intelligence. Miera's map was a prototype. From it, other mapmakers made facsimiles—likenesses that stayed rather true to the original as regards topography and hydrology but deviated here and there in terms of ornamentation. On one, we see the addition of color washes that add drama and beauty; on others, well-meaning corrections to place names misspelled on the original. In the hands of early-twentieth-century borderlands historian Herbert E. Bolton, Miera's map was again modified to influence yet another generation of map users.[14]

Regardless of which copy we consult, the Miera map is a watershed in the explorers' relationship to the interior. With its initial preparation based on the expedition of 1776, and the subsequent but similar copies produced in 1777-78, an important threshold was crossed in the cartographic history of the Great Basin. Wheat observes that Miera's map is an "exploratory *tour de force*" that "represented the first attempt of any European to portray cartographically, from personal experience, the complex upper Colorado River basin, and it was the first to show the lakes and streams of the east-central portion of the Great Basin."[15] It marked the first time that those who actually *experienced the region* began to make maps of it. Before that time, geographic information was provided by informants, then drawn by others. Now, with aggressive reconnaissance, maps are begun in the field, or are at least based on the compilation of field notes. The result is of profound importance psychologically: Such maps are now articulated from the inside out, rather than from the outside in. From about this point on, expectations would never be the same. The eye demands information—the more credible the better—about what lies in the interior. No longer would a simple evocative name like "Quivira" suffice, at least not for the Spaniards who had now experienced the Interior West firsthand. From this point onward, the crown demanded increasingly accurate information based on field experience.

That, however, would be still longer in coming, for mapmaking traditions do not change overnight. Despite a growing interest in accuracy, lack of geographic information in the late eighteenth century did not stop Europeans from mapping fanciful features in the western interior. Rather, it actually *encouraged* cartographers to depict those features that were said, by unnamed or long dead authorities, to exist. Consider, for example, Jonathan Carver's *A New Map of North America from the Latest Discoveries* (1778). Carver depicts a transverse-trending range of mountains as well as a north-south-trending range. With authority, he labels the east-west range "Stony Mountains" and places them between "New Albion" on the west and "Teguayo & Quivira" on the east. To the north of the Interior West, Carver also shows the "River of the West" and "Straights of Anian"—clear references to nebulous legends of European explorers who hoped to find water routes across the continent and wealthy empires within it.[16]

British cartographers were not alone in their beliefs that the interior might be easily traversed and offer new civilizations to discover. Spain claimed this entire region, but the British also vied for a piece of it—New Albion on their maps—which would give them control of western North America and a doorstep to lucrative Asian trade. For her part, Spain also deluded herself about the Interior West, and actions spoke louder than words—or lines on maps. Spain was increasingly knowledgeable about the Interior West's geography, but no map adequately captured its immensity or the challenges that would be faced in crossing it. This region was in fact marginal to Spain as other concerns continued to mount. Although hopes still burned that the interior would prove to be traversed by waterways to California—and that it might possess large and wealthy settlements worthy of Spanish conquest—those hopes were dimming as Spain found itself challenged both internally and externally. Politics and geography worked hand in hand to keep this region isolated: Until the end of Spanish rule in North America (1821), the Great Basin would remain the least explored, and the most poorly understood, of Spain's possessions in the New World.

# 4

## In the Path of Westward Expansion
### 1795–1825

Sometimes maps reveal leaps in knowledge or leaps of faith. Among the most remarkable depictions of the Interior West is that found on William Winterbotham's 1795 *North America* (fig. 4.1). This fascinating map appeared in Winterbotham's popular *American Atlas*. At this time, demand had increased for such books of bound-together maps that featured portions of the world in considerable detail. Winterbotham's *North America* represents a milestone in popular printed maps. Unlike any other map of this period, it clearly shows a prominent north-south mountain range (apparently the Sierra Nevada) and two interior lakes toward which rivers flow. The largest lake appears to be the Great Salt Lake, although it is somewhat farther north on the map than we know it to be. How could Winterbotham depict these geographic features when explorers had not actually traveled this deep into the region? The answer again appears to be Indian informants, who may have communicated with those intrepid Europeans who explored adjacent areas like the Pacific Northwest. The Scottish adventurer Sir Alexander MacKenzie (1764-1820) is a likely candidate, for he reached the Northwest by land in 1793, although people in the United States are incorrectly, if patriotically, prone to credit Lewis and Clark (1804-06) as the first to have done so.

This period is one of great scientific interest and ambitious, geopolitically motivated scientific exploration. By the early 1800s, virtually every European exploration either had a scientific focus or, at least, included sci-

FIGURE 4.1.

W. Winterbotham, *North America* (1795).
Courtesy Special Collections Division, University of Texas at Arlington Libraries

entists on board who made note of the flora, fauna, and other aspects of natural history. The 1791-92 Malaspina expedition to the Northwest coast was
among the more important of such missions. Its immediate goal was to better understand the character of places through expanded description and
inventory of flora, fauna, and landscapes, but its ultimate objective was political control of the frontier.[1] Although early scientific expeditions also tended to hug the coasts, scientists realized that the interiors of continents had
to be explored and documented before any political claims could be vali-
dated. And yet, despite intense and growing scientific interest, Interior
Western North America was still virtually unknown to northern Europeans
who laid claim to large portions of it. Consider the popular book *Voyage à
la Louisiane, et sur le continent de l'Amérique Septentrionale* (Paris, 1802), which

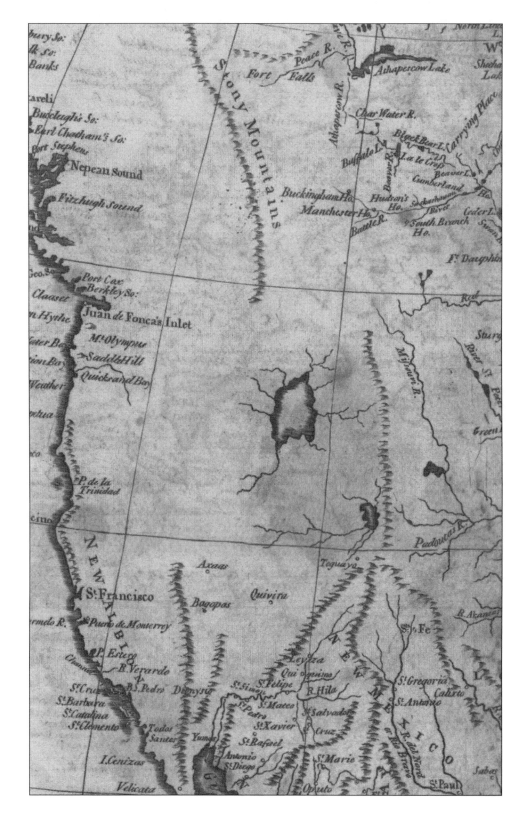

DETAIL OF FIGURE 4.1

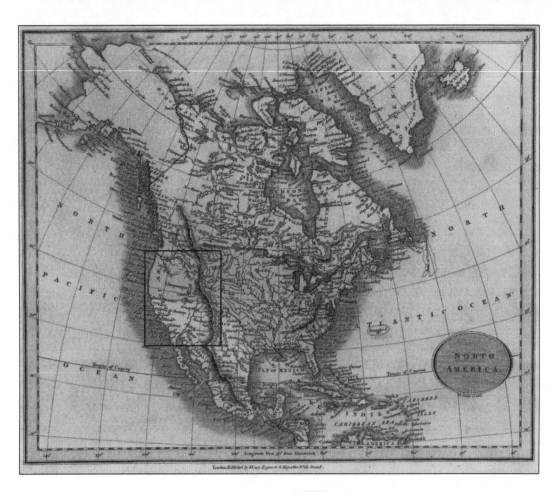

FIGURE 4.2.

John Cary, *A New Map of North America from the Latest Authorities* (1806).
Courtesy David Rumsey Collection

featured a map of *Louisiane et Pays Voisins* by L. Collin. West of a striking
north-south-trending chain of mountains that divides the rivers flowing to
the Atlantic from those flowing to the Pacific, Collin indicates the Interior
West as largely blank space. His words "l'intérieur de cette vaste Contirée
n'est pas encore connu" (the interior of this vast county is not yet known)
are both humbling and accurate.[2] Consider, too, the *New Map of North
America from the Latest Authorities* (1806) by British cartographer John Cary
(fig. 4.2). Published in London, Cary's map delineates the coastlines and elu-
cidates the eastern half of North America in considerable detail. In the
West, however, Cary depicts a virtually complete blank between "L[ake]
Timpanogos L. Sal" and New Albion. That stunning white space—a huge

DETAIL OF FIGURE 4.2

cartographic silence, as it were—must have kindled the imagination. Cary showed considerable restraint and evidently used only verified information. His leaving the interior blank suggests a clean slate, as it were, but in fact other mapmakers were less hesitant in speculating about what lay between the Stony (i.e., Rocky) Mountains and the Coast Range.

Although Cary's map candidly portrayed the interior of western North

FIGURE 4.3.

Alexander von Humboldt, *Map of the Kingdom of New Spain* (1809).
Courtesy Special Collections Division, University of Texas at Arlington Libraries

America as a blank, things were about to change rapidly as Europeans and
Americans intensified efforts to decipher the West. Among the crucial events
in North American cartography was the visit to New Spain by Alexander
von Humboldt in the late 1790s. This scientific and diplomatic visit culmi-
nated in the publication of Humboldt's *Map of New Spain* (fig. 4.3) in 1809.

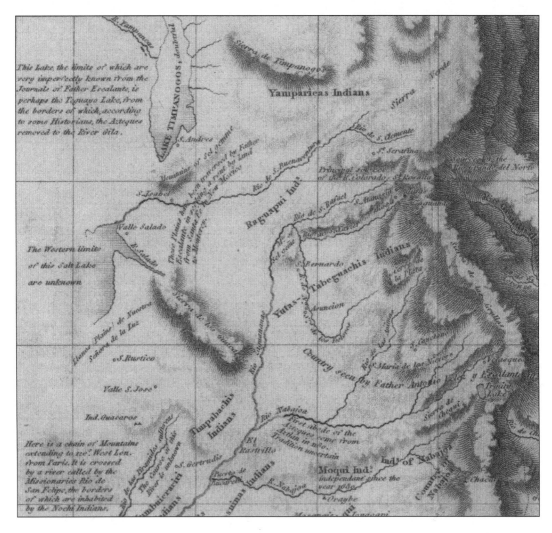

This Lake, the limits of which are very imperfectly known from the Journals of Father Escalante, is perhaps the Timpano Lake, from the borders of which, according to some Historians, the Aztecs removed to the River Gila.

The Western limits of this Salt Lake are unknown

Here is a chain of Mountains extending to ye West Lon. from Paris. It is crossed by a river called by the Missionaries Rio de San Felipe, the borders of which are inhabited by the Nochi Indians.

LAKE TIMPANOGOS, or general

Yamparicas Indians

Raguapui Ind.

Tabeguachis Indians

Yutas

Country seen by Father Antonio

Timpabachis Indians

Moqui Ind.s independent since the year 1680

Ind.s of Navajo

DETAIL OF FIGURE 4.3

Humboldt's original *Map of the Kingdom of New Spain,* hand-drawn in 1803, represents the state of knowledge at the time. The map's northwest corner is of special interest as it depicts a portion of the Great Basin. It clearly shows that Humboldt had consulted maps by Miera and other early explorers, which likely resided in Mexico City's archives. Humboldt himself refers to the Escalante expedition on his map and uses essentially the same framing of the "Lac de Timp" at its northwest. Note, too, that a river flowing east to west suggests the influence of the 1777 Miera map (or maps that were derived from Miera's) on Humboldt's. Based on as many sources as he could locate, Humboldt endeavored to articulate the interior of Upper California, as the area was now called, but knew his limits. Hinting at the conflicting information that confronted him, Humboldt declared that "I chose rather to leave vacant space in my map than to draw from suspicious sources."[3]

Although striving for accuracy, Humboldt nevertheless reproduced several features, like the Rio Buena Ventura, that were to prove vexing to mapmakers and intriguing to politicians. Being a scientist rather than speculator, Humboldt indicated hydrographic features only tentatively. Interestingly, however, he did yield to a bit of speculation when he noted that the site of the Aztecs' early development might be located in the Intermountain West. That claim would stimulate generations of anthropologists and help lead to the creation of a potent, if mythical, Aztlán in the American Southwest.

We can learn much about Humboldt—and his difficulty in mapping the interior portion of North America—from his insightful *Political Essay on the Kingdom of New Spain.* In it, Humboldt revealed the difficulty he had in mapping portions of northern New Spain. "I could only give a very imperfect map of Mexico," he admitted, adding that "[i]t will be said, without doubt, that it is yet too soon to draw up general maps of a vast kingdom for which exact data are wanting."[4] This problem of incomplete data would plague Humboldt as it had other cartographers. Nevertheless, he persevered, consulting existing maps wherever possible, including the *Mapa del Nuevo Mexico,* which, he observed, "is very minute [i.e., detailed] with regard to the countries situated under the parallel of 41°" and "contains details as to the lake *des Timpanogos.*"[5] Humboldt did not further identify the cartographer who made this map, but it evidently influenced him.

But it was Humboldt's own map that was now in the limelight. He asserted that "notwithstanding great imperfections, my general map of New Spain has two essential advantages over all those which have hitherto appeared"—namely, it documented the locations of important mines and indicated "the new division of the country into intendancies," or jurisdictions. But physical geography was also Humboldt's passion, and it was here that he recognized his map's shortcomings. He admitted that "indication of the chains of mountains presented difficulties which can only be felt by those who have been themselves employed in constructing geographical maps." To convey the topography most effectively, Humboldt used a combination of two projections so that the relief, or profile, of the mountains could be represented. He admitted that this method, "the oldest and most imperfect of all [solutions], occasions a mixture of two sorts of very heterogeneous projections." By "projections," Humboldt means techniques that can help the map user visualize the lay of the land, rather than the meaning of the term that we understand today. He stated that he used hatchings, or hachures, to accentuate the topography of the mountains. These lines drawn to suggest the topography coincided with a growing interest in describing significant features, such as mountains, in considerable detail. Seemingly innocuous, those hachures actually had the power to disguise deficiencies in topographic knowledge. Humboldt himself acknowledged as much when he noted that "hatching . . . forces the drawer to say more than he knows, more than is even possible to know of the geological constitution of a vast extent of territory."[6]

Humboldt's words and maps reveal a tortured process of self-examination—a scientist encountering and evaluating a wealth of information, some accurate, some inaccurate, and making profound decisions about what to trust and what to discard. By deduction, Humboldt attempted to produce the finest map available of New Spain—one that was both accurate and useful to scientists as well as potential resource users. He took pains to explain the most effective way of representing a wealth of physical and cultural information. His realization that "the waters undoubtedly in some sort give delineation to the country," is prescient, but it was hydrology that ultimately tricked him. As Humboldt himself observed, "A knowledge of the great vallies or of the basins; an examination of the points where rivers take their rise; are certainly extremely interesting to a hydrographical engineer; but it is a false application of the principles of hydrography, when geographers attempt to determine the chains of mountains in countries of which they suppose they know the course of rivers. . . . Hence the attempts which have been hitherto made to construct physical maps from theoretical ideas have never been very successful."[7]

Humboldt was essentially correct that theory without field observation is risky. However, it was precisely the discipline he warned about—theoretical hydrology—that would ultimately help mapmakers deduce the internal drainage of the Great Basin. Observers needed to know not only *where* watersheds existed, but *what happened to the waters* once they began to move downward from mountain crests. Despite his astuteness and genius, Humboldt would not unlock the riddle of the Interior West, nor would anyone until hydrology and topography were wed a generation later.

Yet another factor stood in Humboldt's way, and that was sheer distance. In his defense, Humboldt never personally traversed this region, only representing it on his map as carefully, and with considerable political skill, as his friendly relations to Spain would permit. Nevertheless, his *Map of New Spain* is significant for several reasons. First, it generated considerable debate in his own time, figuring in one of the great cartographic controversies of the early nineteenth century. The controversy confirms that science involves claims and counterclaims of authorship. It also confirms that the Age of Science was one of personal ambition played out on a geopolitical stage. Mapmakers could no longer borrow from each other without concerns about plagiarism arising. In publishing his map of New Spain, Humboldt was careful to cite numerous Spanish maps as well as the *Map of Louisiana* published at Philadelphia in 1803. Others, however, were not as cautious or considerate. If scientific findings demanded both authorship and documentation, Humboldt was prepared to fight for them, especially when his work was pirated. The controversy centered on the relationship between Humboldt's map and Zebulon Pike's much-debated *Map of the Internal Provinces of New Spain* (fig. 4.4). Pike had been selected by Jefferson to help resolve a burning issue—the western border of the Louisiana Purchase. Lewis and Clark helped to clarify the Northwest, but the southwestern border was in dispute from the very beginning. Once again, water proved to be an important element in conceptions and misconceptions of territory. Spain claimed that the Red River marked the United States-Spanish border, while Jefferson claimed the Rio Grande. All of this political maneuvering took place far from the Great Basin; yet Pike sensed and even led the national mood for westward expansion. Not so coincidentally, Pike's map extends far to the west of the disputed area, including the eastern portion of the Great Basin—exactly the area, notably the eastern portion of the Great Basin, that frames the northwestern corner of Humboldt's map. The fact that both maps feature Lake Timpanogos (Salt Lake) at their far northwestern corner is not coincidental, but very revealing. Like Humboldt's original map, Pike's map also features the known (and conjectural) hydrographic landmarks in exactly this region. Pike's map is so similar to Humboldt's that an impartial observer can readily see that one map is clearly derived from the other. Humboldt claimed that Pike had copied his map, and that claim is plausible. After all, Humboldt had left a copy of his original map with Thomas

Jefferson when he visited Washington in 1804. Pike had also visited Washington but, for his part, claimed that his map was an original based on field experience. Pike, it should be noted, was a master spy who might have seen (and copied) the same Spanish maps that Humboldt himself had used. The intrigue does not end there: British cartographer Aaron Arrowsmith was also involved, for he, too, had produced a similar map!

The appearance of maps similar to his *Map of New Spain* perturbed Humboldt, who made public his disapproval of Pike's apparent plagiarism. Within a few months, Pike and Humboldt were center stage in what had become an international cartographic controversy. In correspondence to "my dear Baron," Thomas Jefferson confirmed that he had received several of Humboldt's volumes and "an interesting map of New Spain" which "give us a knoledge [sic] of the country more accurate than I believe we possess of Europe." But Jefferson then added that "I fear from some expression in your letter, that your personal interests have not been duly protected, while you were devoting your time, talents & labor for the information of mankind."[8] That concern was about the map that, Humboldt claimed, had been used without permission or even acknowledgment by two parties—Pike and Arrowsmith. More than a hint of Humboldt's anger is evident in a subsequent letter to Jefferson, in which Humboldt charged that "Mr. Arrowsmith in London has stolen my large map of Mexico, and Mr. Pike has taken, rather ungraciously, my report which he undoubtedly obtained in Washington with the copy of this map, and besides, he also extracted from it all the names. I am sorry over my cause for complaint about a citizen of the United States who otherwise showed such fine courage. I don't find my name in his book and a quick glance at Mr. Pike's map may prove to you from where he got it."

Ever the Anglophobe, Jefferson responded to Humboldt on December 6, 1813. Citing Arrowsmith's "Anglo-mercantile cupidity," Jefferson enlarged upon his meaning: "That their Arrowsmith should have stolen your map of Mexico, was in the piratical spirit of his country." If this made von Humboldt feel a bit better, Jefferson's next words were meant to steel him for a defense of Pike, who had been killed in battle in Canada: "But I should be sincerely sorry if our Pike has made ungenerous use of your candid communication here; and the more so as he died in the arms of victory gained over enemies of this country." Lest Humboldt misunderstand this diplomat-

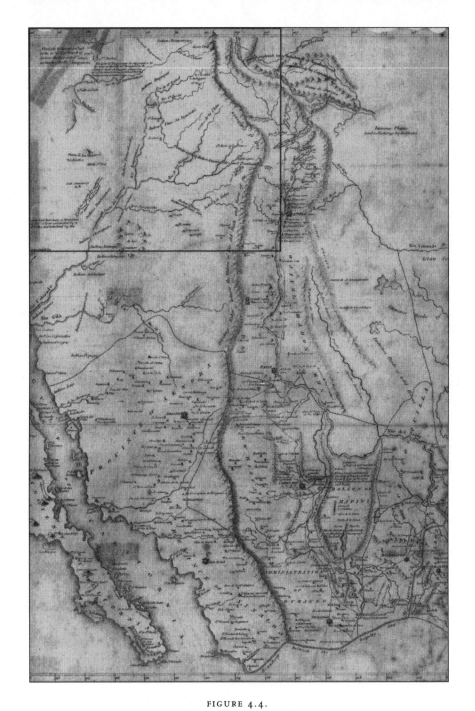

FIGURE 4.4.

Zebulon Pike, *Map of the Internal Provinces of New Spain* (1810).
Courtesy Special Collections Division, University of Texas at Arlington Libraries

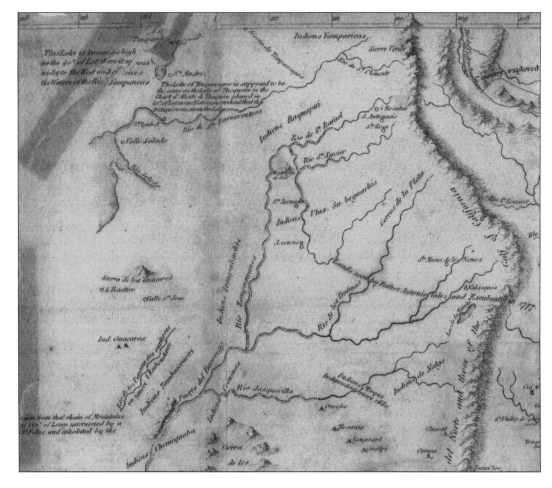

DETAIL OF FIGURE 4.4

ic language, Jefferson elaborated: "Whatever he [Pike] did was on a princi-
ple of enlarging knolege [sic] and not for filthy shillings and pence" but
rather "to excite an appeal in his readers." Hoping to end the matter there,
the former president added: "I am sorry he omitted even to acknolege [sic]
the source of his information. It has been an oversight, and not at all in the
spirit of his generous nature."[9]

Despite any similarities, borrowed or otherwise, between maps by Humboldt, Pike, and Arrowsmith, they served somewhat different purposes.
Humboldt's was meant to inform the world about the richness of opportunities and resources in New Spain, an appreciation of the graciousness of the
Spaniards who hosted, and toasted, the baron. These were tense times
indeed. Spain was furious at the French for selling Louisiana to the United
States, for that brought the expansion-minded young country right to its

doorstep. Thomas Jefferson's sending Lewis and Clark to find a water passage through the continent had been greeted with both suspicion and outright consternation by Spain.[10] Spain realized that Pike's map played a less gracious and more ambitious role than Humboldt's. Pike's widely produced work served as an enticement to Anglo-Americans who were anxious to learn more about the wealth of New Spain and its possible inclusion into an expanding young nation. And Arrowsmith's map served to inform the British of lands lying to the south of their claimed area in the Northwest, namely today's Oregon and Washington. With luck, Spain's presence would continue to check American expansion into England's eroding North American empire, an action that would guarantee new trade and mercantile opportunities for British companies.

National expansion called for imaginative mapmaking, and not even mountains were immune to being manipulated. On their influential map *Louisiana,* which appeared in the *New & Elegant Atlas* in 1804, Arrowsmith and his fellow cartographer Samuel Lewis had essentially eliminated the Intermountain West by drawing the spine of the Rocky Mountains—labeled jointly in French and English as "Mn de la Roche or Stoney Mtns"—far to the west of where the range actually existed. Interestingly, here Thomas Jefferson's information, or rather misinformation, may have tricked the British mapmakers. Arrowsmith and Lewis perpetuated a geographic myth that had been "advanced persuasively by a number of Jefferson's most respected scientific colleagues in North America and Europe." In this geographic mythology, the Rocky Mountains "bear about the same relationship to the Pacific as the Blue Ridge does to the Atlantic."[11] Even Winterbotham's 1795 map of North America features this mythical nexus of mountain ranges. In one fell swoop, topography (mountains) and hydrography (watershed) conspired to influence a generation into downplaying physical obstacles and to envision an easy jumping-off place downslope to the Pacific. In Jefferson's time, there was widespread belief in a "core drainage area, or 'pyramidal height-of-land'" from which all western rivers flowed to the Pacific. To quote late-eighteenth-century minister James Maury, who believed that Providence played a role in configuring the land, it would require but a simple portage "across which a short and easy communication opens itself to the navigator."[12] With considerable wishful thinking, maps depicted one simple crest dividing the waters of the continent, and they significantly placed this crest very close to the Pacific. In reality, the confusion stemmed from the paucity of actual exploration in this part of the West, but it conveniently, if not incredibly, disposed of almost two hundred thousand square miles of vexing topography—the Great Basin—by cartographic sleight of hand. Instead of a punishing interior that had thwarted Spaniards' efforts to reach the Pacific by an overland route a generation earlier, Jefferson persuasively abolished that interior, cartographically speaking. That erasure helped suggest and endorse the inevitability of American expansion to the Pacific and helped fuel British fears of that American incursion into the Northwest. For their part, perceptive Spaniards began to sense that their country's days of hegemony in the Southwest might be nearing an end.

By about 1810, British and French interest in the Far West was becoming insatiable as the region suggested both opportunity and adventure. Hubbard Lester's popular book *The Travels of Capts. Lewis & Clark* featured a map by Samuel John Neele entitled *Map of the Country Inhabited by the Western Tribes of Indians.* Neele's work is based on earlier maps by Samuel Lewis and Antoine Soulard and places a north-south-trending range of mountains (what would later be called the Rocky Mountains) far west of where they actually exist.[13] An almost identical map, but with French inscriptions, appeared in Patrick Gass's popular *Voyage des Capitaines Lewis et Clarke* (1810).[14]

It was inevitable that maps by Humboldt, Pike, Lewis, Soulard, and Arrowsmith would inspire lesser cartographers to churn out copies in the 1810s and 1820s. We might call these products "knockoffs" today, but imitation was justified as a form of flattery and proved a good source of income. One very popular version, *Amérique du Nord,* was printed in Paris and is identical to the map in Humboldt's famous French atlas of the 1810s. Those maps were popular because they helped entrepreneurs and politicians visualize empires. Maps of the early nineteenth century contained geopolitically motivated geographic features that might encourage both trade and expansion by those so inclined. Although both Humboldt's and Pike's maps do not indicate the full course of a westward-flowing river across the Interior West, increasingly fervent American narratives suggested that it ran all the way to the Pacific. This river was long a dream of Spaniards, but consider for a moment the significance that it would now possess for Americans. This westward-flowing river would seem, in fact, to endorse

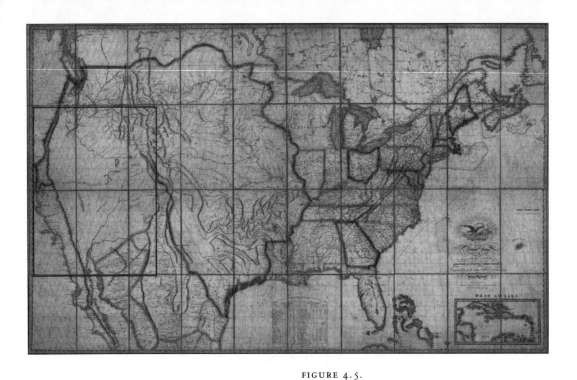

FIGURE 4.5.

John Melish, *Map of the United States . . . from A Geographical Description of the United States, with the Contiguous British and Spanish Possessions* (1815). Courtesy David Rumsey Collection

Americans' "natural" inclinations to reach the West Coast by traveling through the interior. In developing suppositions about water routes across the continent, proponents of western expansion increasingly relied on maps. One of the most widely consulted maps and reports of the period, John Melish's *Map of the United States with the Contiguous British and Spanish Possessions Compiled from the Latest and Best Authorities* (1816),[15] positions two western rivers as conduits of expansion (fig. 4.5). These rivers—the Multnomah and Buenaventura—are depicted as cutting through the Interior West directly to the Pacific. "The Multnomah," Melish noted, "is supposed to rise near the head waters of the Rio del Norte [i.e., Rio Grande]. . . . Viewing it in its connection with the head waters of the Missouri, the La Platte, the Arkansas, and the Rio del Norte, it deserves particular notice, as it will probably be, at no very distant period, the route of an overland communication, through the interior of Louisiana, to the settlements at the mouth of the Columbia river."[16] In this assessment, Melish evidently built on earlier maps that show the Multnomah's ambitious diagonal course

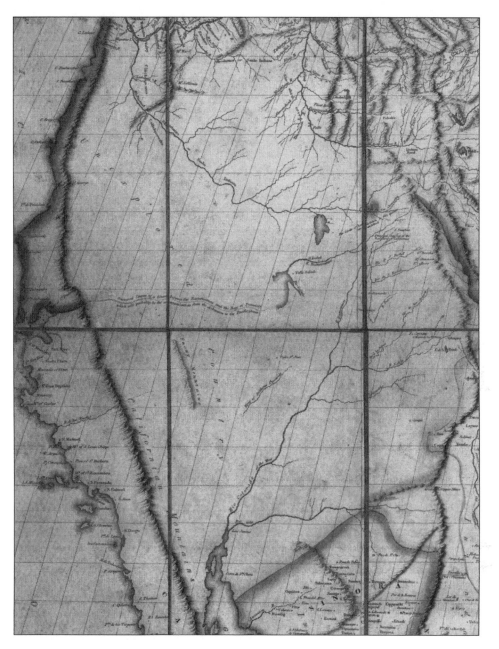

DETAIL OF FIGURE 4.5

through the Great Basin. Only later would exploration reveal the Mult-nomah's more restricted course: with headwaters in the Cascades, it is con-fined entirely to the Pacific watershed in Oregon.

Melish was even more certain of the Rio Buenaventura's potential for east-west commerce. As he put it, "There is little doubt, therefore, but that the Rio Buenaventura, and its waters, which interlock with waters of the

FIGURE 4.6.

John Robinson, *Map of Mexico, Louisiana, and the Missouri Territory* (1819).
Courtesy Special Collections Division,
University of Texas at Arlington Libraries

Rio del Norte, and La Platte, form part of it. Should this be the case, it may, in process of time, form an admirable communication with the settlements on the west coast of America."[17] Melish's 1816 map labels as "unexplored country" the land between the Rockies and the Sierra Nevada but again succumbs to the popular conceptions of the times by indicating the course of a river between the Great Salt Lake and the Pacific with the following evocative prose: "Supposed course of a River between the Buenaventura and the Bay of Francisco which will probably be the communication from the Arkansaw [*sic*] to the Pacific Ocean."[18] Scottish-born but becoming a citizen of the United States by choice in the early 1800s, Melish shared his vision of westward expansion through the maps he produced in his adopted city of Philadelphia. Robert Mills, one of America's strongest advocates

DETAIL OF FIGURE 4.6

of early transportation improvements, included Melish's map in his major treatise of inland navigation's role in reaching the Pacific Coast.[19]

Other cartographers were equally enthusiastic about the potential of rivers in the Interior West to aid in westward expansion. Doctor John Robinson's 1819 *Map of Mexico, Louisiana, and the Missouri Territory* (fig. 4.6) includes several mythical features, among them rivers flowing from the vicinity of Lake Timpanogos (the Great Salt Lake) all the way to the Pacific. Robinson was an outspoken proponent of American expansion and had accompanied Pike on his explorations in the West. On Robinson's map, no fewer than *three* rivers flow into the Pacific from the Interior West. Through such enthusiastic interpretations, one river in particular, the fabled Rio Buenaventura, became a fixture on maps in the early nineteenth century. This river is especially evident on popular maps depicting the Missouri Territory, and its inclusion here confirms its seamless integration into the narrative, and imagery, of American westward expansion. It is but one of many mythical geographical features that found their way

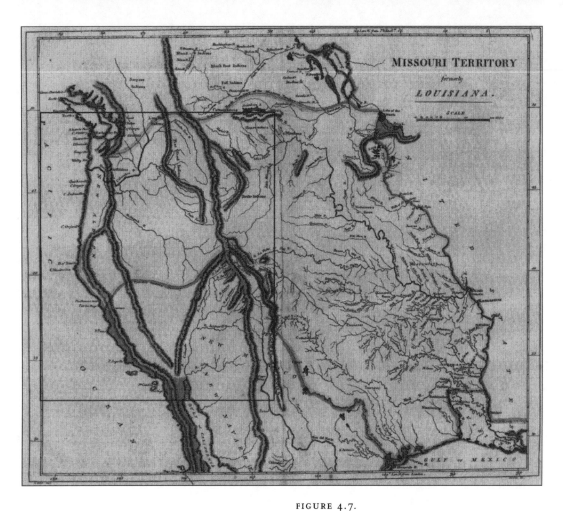

FIGURE 4.7.

John Bower, *Missouri Territory, Formerly Louisiana* (1814).
Courtesy Special Collections Division, University of Texas at Arlington Libraries

onto maps in a process that Seymour Schwartz calls "the mismapping of America."[20]

American cartographers like Melish and Robinson built on maps by Spanish, French, and British mapmakers, as well as hearsay from explorers. By the mid-1810s, American cartographers had made considerable progress in delineating the Interior West, but many misconceptions still remained. Typical of the efforts of this period, John Bower's 1814 *Missouri Territory, Formerly Louisiana* (fig. 4.7) reveals the cartographer's desire to delineate the courses of rivers well inland (including the persistent Multnomah, which drains a large portion of the Interior West). On it, the mountains clearly elude the cartographer, reminding us that the region could only be under-

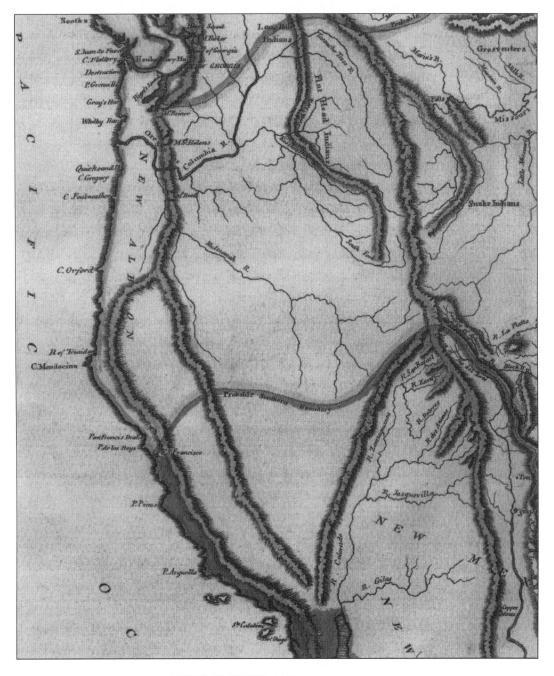

DETAIL OF FIGURE 4.7

stood when two physical elements—hydrology and topography—were coordinated. Bower's map is very similar to other maps depicting the Missouri Territory, including one by Samuel Lewis that was also popular in the period 1815-1820. Consider the pressures that faced both explorers and mapmakers at this time. It was one thing for explorers to traverse regions, quite another to accurately survey those regions. Accurate delineation of a region is as much a political process as an adventure in the field. Historian D. Graham Burnett noted one sobering fact about surveys into otherwise exotic *terrae incognitae:* "Reaching the right sort of unknown regions demanded that an explorer negotiate committees, history, and budgets as well as rivers. But getting to someplace incognita did not complete the task. The obligation of a geographical explorer was to return from a cartographic blank with a map."[21]

Then, too, consider the political ramifications of every mountain range, every watershed, and the pressures are compounded. The explorer may seem independent to us as we romanticize his accomplishments. But, like mapmakers themselves, explorers have sponsors whose expectations are never very far away despite the immense distance traversed. The interior North American West may seem to isolate the explorer from his sponsor, but it could not insulate him from expectations of the crown or president. Anthropologist Françoise Weil observes that the explorer is motivated by two sets of perceptions, his own agenda and the agenda of his backers— those "silent partners" who not only enable him to explore but also have to "see through the account" that he makes.[22]

Maps of the interior American West at this time reflect a myriad of powerful, sometimes conflicting, forces. They are also based on geographic information that is the result of conflicting perceptions, motivations, and observations. If it is difficult to find two maps depicting exactly the same features, that is because imaginations were both fertile and fickle. Historian of exploration John Logan Allen observes that "the rivers of wonder and mountains of myth . . . were fixed neither in maps or minds and remained as geographical features that wandered about the western landscape."[23] Small wonder that maps portray some bewildering oddments. Yet despite these maps' errors, some of which might involve wishful thinking or political manipulation, one can see that a spatial differentiation was occurring as regards the geography of the interior. For its part, even the mythical Rio

Buenaventura is not purely political chimera. It does have a counterpart that makes, at least in part, the journey shown on maps. Today's Humboldt River does indeed flow westward as it crosses much of the region; it simply does not flow as far as observers thought. The problem lies not in the hydrology—for a river does indeed flow westward across much of the Great Basin, and a river does indeed flow westward into the central/northern California coast at about the same latitude. Rather, the problem lies in incomplete observation. It reflects a lack of information about another geographic feature—the Sierra Nevada—that lay between the interior and the coast.

Recall, too, that the presence of a mountain range might not dictate that a river had to change course. In some places rivers flow *through* mountain ranges, as evident in the "water gaps" of the eastern United States. If rivers breached some of the more rugged country of the East, why not the West, too? And, indeed, some rivers in the West *do* breach mountains. Consider, for example, the Columbia River, which flows through the imposing Cascade range east of today's Portland, Oregon. In California, the Sacramento/San Joaquin river system breaches the Coast Range to empty into San Francisco Bay. At this time, most Anglo-American cartographers seemed largely unaware of the Sierra Nevada defining the western edge of the Great Basin region. Believing that a river flowed through the coastal mountains at about the same latitude, most cartographers simply drew the river as they thought it *should* run, which is to say all the way to the coast, by connecting the dots, as it were. If they did not understand the region's hydrology, that was in part because their assumptions outran their experience. In other words, more information was needed before the region could be articulated according to its actual—which is to say unusual, internally drained—hydrography.

No chapter on the early-nineteenth-century cartographic history of the American West would be complete without reference to one of the region's most enigmatic and underappreciated mapmakers—Juan Pedro Walker. Born in Spanish Louisiana in 1781, the enterprising, trilingual Walker made maps of the region in the early 1800s. Historian Elizabeth John suggests that Walker's maps likely influenced Pike and Robinson.[24] Walker's maps of the Southwest, especially Texas, Chihuahua, and Coahuila, are fairly well documented. However, one of his efforts—an untitled map depicting the western portion of North America from the Great Lakes to the Pacific (fig.

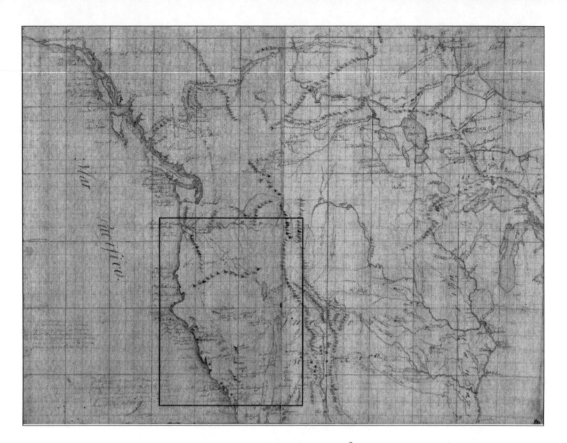

FIGURE 4.8.

Juan Pedro Walker's untitled manuscript map showing the western portion of North America (ca. 1817). Courtesy The Huntington Library, San Marino, CA

4.8)—is breathtaking in its sweeping coverage of the region. "In many respects the most ambitious of all" of Walker's maps,[25] it contains some surprisingly sophisticated information for its early date (which cartographer Carl Wheat assumed to be 1817).[26] As a Spanish citizen and mapmaker, Walker parlayed information from his extensive travels. He also would have had access to the most sophisticated maps of the region, and his map confirms this. It articulates the rivers of the Great Basin region in more detail than any other map of the period.

It is here that Spanish ambitions and Spanish secrecy come into play, cartographically speaking. Walker's map reveals that Spain knew far more about this region than she was willing to divulge to the public. One can only envision the anguish Spain felt as Anglo-American and British cartographers like Pike and Arrowsmith began delineating her northern frontier with greater and greater accuracy. Even the errors—the suggestion of an easy

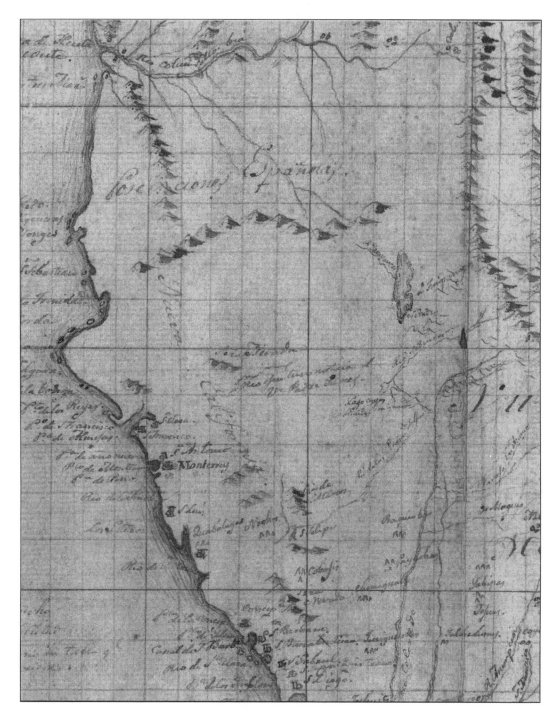

DETAIL OF FIGURE 4.8

route across the Interior West that had enchanted Spain a century earlier—now seemed a menacing metaphor for the relentlessness of the Anglo-Americans' impulsive westward expansion. When cartographic historian Robert Martin called Dr. Robinson's map a blueprint for revolution, he was correct. As keys to the West, maps now became weapons in the hands of expansionists heeding the early call of an impulse that would be called Manifest Destiny by the next generation.

# Demystifying *Terra Incognita*
## 1825–1850

The early-to-mid-nineteenth century marked a time when cartographers scrambled to try to depict the topography and hydrology of the West accurately. Despite their efforts, however, considerable speculation was common, and errors frequent. Consider Anthony Finley's interesting but conflicted *Map of Western America* (fig. 5.1). Produced in 1826, it shows a fragmented Sierra Nevada breached by numerous rivers. Finley's disjointed mountain range obediently parts to permit the westward-flowing rivers of empire to reach the Pacific in California. This map reveals a conundrum that would soon be resolved, and that resolution reminds us that cartography is a complex drama featuring many actors. Some of these actors were explorers whose experiences confronted widely held beliefs, long-cherished cartographic traditions, and deeply rooted cartographic icons.

Consider, too, the case of those who longed to make sense out of the eastern portion of the Great Basin. The key here would be determining the shape of what we today call the Great Salt Lake, which was circumnavigated by a party of American explorers under General Ashley as early as 1826. Like other observers, they were well aware of the lake's hypersalinity. To us with the advantage of hindsight, the salinity of the Great Salt Lake seems like a simple, definitive clue to the region's aridity. After all, we now know that evaporation exceeds precipitation, and that this remnant of an inland sea (Lake Bonneville) increased in salinity as it shrank in size. But the very factor of its salinity worked against explorers' understanding its real charac-

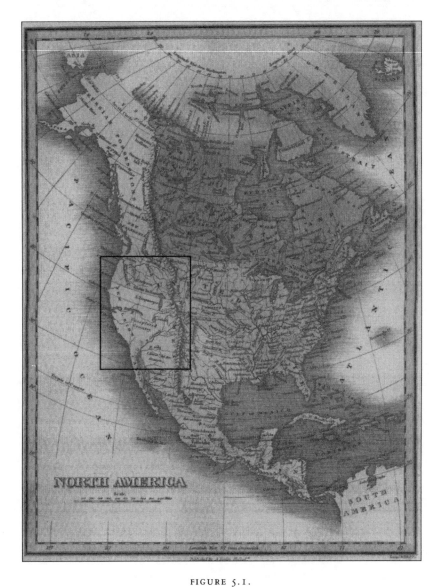

FIGURE 5.1.

Alexander Finley, *North America* (1830). Courtesy Special Collections Division,
University of Texas at Arlington Libraries

ter. When they first encountered the Great Salt Lake, they were amazed by
its "bitterness," but that only seemed to prove that it was an arm of the sea,
for what else could a body of salt water this large be? Beliefs like this die
hard, even in the face of evidence. In 1826, *Niles' Register* reported that
Ashley's men had coasted the perimeter of the lake. Evidently, however, the
explorers were unable to comprehend a lake with no drainage to the sea. So

strong was their faith that they believed their assumptions rather than what they had actually experienced: The reporter noted that although "[t]hey did not exactly ascertain its outlet[,]" they felt certain that they must have "passed a place where they supposed it must have been."[1]

These explorers were deep in the wild heart of interior Northern Mexico at the time. Like others before them, they were seeking the Pacific unaware of the Sierra Nevada's stranglehold on rivers. Throughout the 1820s and early 1830s, the interior of Alta California, including much of today's Nevada and Utah, would remain largely unknown, even though an established trade route now skirted its southern edge. But the region's isolated status actually invited new interest by others wishing to cross it as they sought the Pacific ports and trade with the Orient. These Anglo-Americans relied on maps from the early 1800s that tended to simplify the topography and grossly distort distances. Still revealing influences of the Jeffersonian-era tendency to misrepresent and misplace the Rocky Mountains, these maps severely *compress* the distance between Pike's Peak and the Pacific Coast. For example, although the distance from Pike's Peak to the Pacific is really 1,006 miles, Robinson's (and later) maps show it to be but 660 miles. Consider this compression a metaphor: It effectively squeezed the Intermountain West out of the public's consciousness. The promoters of western railroads to the Pacific Ocean in the 1820s and 1830s were especially prone to downplay the great distances and formidable topography that would be encountered here.[2] This oversight by mapmakers was understandable for two reasons. First, relatively little information was available. Second, few obstacles and short distances further encouraged belief in technology's ability to span the West. The mapmakers' combined ignorance and wishful thinking helped support their advocacy of a railroad line across the region fully four decades *before* the transcontinental railroad ultimately crossed it in 1869.

Ignorance was inadvertently assisted by the cartographers who perpetuated errors from previous maps. Such geographic oversights were understandable, for simplification was part of the process of making sense of the region. As historian William Goetzmann observes, "[T]he American West, that vast unknown at the beginning of the nineteenth century, provides a significant case study of the many ways in which explorers made visual interpretations that simplified the complex geography of a place or places that the masses of people would not yet experience for themselves."[3]

On the 1830 *Amérique du Nord* map by A. H. Dufor (Paris) that owes such a debt to Humboldt, the Great Basin shows as a startling blank with two bands of text intersecting. One, running diagonally from southwest to northeast, reads, "TERRITOIRE D'INDIENS INDEPENDANS" (Territory of Independent Indians). These Indians were not confined to reservations and were highly mobile. This map recognizes their presence across a huge portion of Mexico's northern frontier. (Those Indians are called "Freier Indianes" on German maps of the same period.) The other band of text runs from northwest to southeast and reads, "PAYS INCONNU" (Unknown Land). This, of course, is the French equivalent of *terra incognita*. As with most other maps from this period, Lake Timpanogos and other features near the Wasatch Mountains are depicted, and westward-flowing rivers breach the mountains bordering coastal California. Overall, however, one marvels at the mapmaker's restraint. To Dufor, much of the Interior Western North America was literally unknown, and he showed it as such. Humboldt's scientific caution had indeed rubbed off on subsequent mapmakers.

But those open spaces that still resonated as *terra incognita* taunted others less restrained in ambition and rhetoric. At precisely this point in time, *terra incognita* became part of the challenges facing "Young America," a complex sociopolitical development involving: (1) *literary nationalism,* which is to say an American literature that was different from English literature; 2) *Manifest Destiny,* which embodied expansionist sentiments celebrating American power and righteousness; and 3) *modernist capitalism,* which included faith in technology and worldwide trade.[4] It should be noted that literary nationalism had a cartographic analog—the development of American maps prepared in places like Philadelphia, New York, and Boston by individuals who passionately supported American expansion and sought to obliterate blank spaces. The cartouches of these maps often featured American icons, such as the eagle. However, what especially distinguished these maps was their depiction of geographical features—rivers and passes, for example—that would facilitate westward movement. They were, in effect, blueprints for American expansion.

Being part of Mexico's far-flung northern frontier, the Great Basin remained lightly populated and hence vulnerable to intrusion by interlopers who had at least loose allegiances to Anglo America. Anglo-Americans took a much closer look at this region as their dreams of westward territorial expansion took shape. Although largely desert in nature, it did contain diverse mountainous areas and perplexing rivers. The Interior West also possessed rich resources. Increasingly aware of growing markets, mountain men like the legendary Peter Skene Ogden and James Bridger—and trapper-writer-cartographer Warren Angus Ferris—would soon advance geographic

FIGURE 5.2.

Benjamin Bonneville, *Map of the Territory West of the Rocky Mountains* (1837).
Courtesy Special Collections Division, University of Texas at Arlington Libraries

knowledge about the region in ways that began to satisfy both scientists and politicians. Among the most interesting maps of this period was Benjamin Bonneville's 1837 *Map of the Territory West of the Rocky Mountains* (fig. 5.2).

Bonneville's map breaks new ground, a result of his diligent research and ambitious exploration. It shows an impressive mountain range (the Sierra Nevada) defining the western edge of the Interior West, a prominent river running diagonally through the region, and another mountain range running diagonally through the area occupied by the Shoshone Indians. Bonneville's map is prescient, and deserves closer inspection. It deciphers precisely that area that had bewildered the Spaniards half a century earlier and still confounded many Anglo-Americans: On Bonneville's map, the

Great Salt Lake (Lake Bonneville) is separate from the southwestward-flowing river (the "Mary or Ogden's River" on the map) that finds its way to smaller lakes at the base of the Sierra Nevada (called the "California Mountains"). Of the Ogden River, he notes that trappers had "ascertained that it lost itself in a great swampy lake, to which there is no apparent discharge."[5] This, of course, is the fate of the Humboldt River, and Bonneville's map correctly depicts it.

Bonneville's map was also sobering to those who thought crossing the region to the Pacific Coast would be easy. It dispelled the heady notion of a river flowing to the Pacific from the Interior West. Bonneville's maps and writings reveal his fascination with the region's hydrology. He was virtually obsessed with its most spectacular body of water—the Great Salt Lake: "This immense body of water spreading itself and stretching further and further, in one wide and far-reaching expanse, until the eye wearied with continued and strained attention, rests in the blue dimness of distance, upon lofty ranges of mountains, confidently asserted to rise from the bosom of the waters."[6] To the romantic early-nineteenth-century imagination, the region's unusual geographic features and vast distances lent a sense of the sublime. Yet, as science demanded proof or verification, maps served as an intermediary; because strict cartographic conventions had developed, the fanciful or imaginary was more and more constrained by graticules and tight lines. Bonneville's map hints at a sobering reality that was slowly becoming understood—that mountains apparently defined the perimeter of the Interior West. And yet, another river in the southern part of Bonneville's map—what he labels as the "Lost River"—still leaves some hope that a waterway might reach to the California coast south of those mountains. This lost river is historical in that it recalls the enigmatic stream Juan de Oñate was unable to cross more than two centuries earlier. It is also metaphorical in that it signals the last hope of a water route through the Interior West to the Pacific.

By 1838, David Burr's *Map of the United States of North America with Parts of the Adjacent Countries* (fig. 5.3) represented even more refinement. At the time of its publication, the Sierra Nevada was beginning to be comprehended as a stupendous barrier that would bar any rivers in the Interior West from flowing to the Pacific: As the first popular map to delineate the Sierra Nevada range, Burr's now recognized that substantial rivers (like the

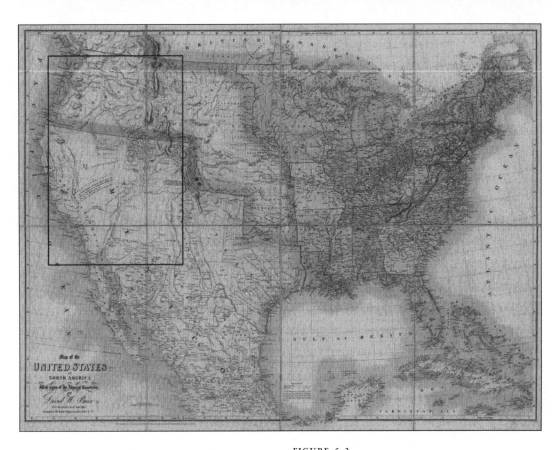

FIGURE 5.3.

David Burr, *Map of Western Portion of the United States of North America* (1838).
Courtesy David Rumsey Collection

American and Sacramento) flow westward *from* those mountains. Burr's
map showed these rivers arising from headwaters in the Sierra Nevada,
rather than running through the range from headwaters in the Great Basin.
Literally and metaphorically, this map is a watershed in the region's cartographic history. Although it definitively states that the Interior West is a
"Great Sandy Plain"—a statement that seems to downplay the Great Basin
region's largely mountainous character[7]—Burr's map comprehended an
important aspect of the region's peculiar hydrology. On the map itself, Burr
notes that the region's rivers and streams are "soon absorbed in the sands."
On his revised 1839 map of the same title, Burr again depicts the Great
Basin portion of the Intermountain West as a "Great Sandy Plain," but rather
than give the impression of one monotonous flatland, he adds an explana-
tory note: "Some isolated Mountains rise from the Plain of Sand." Streams

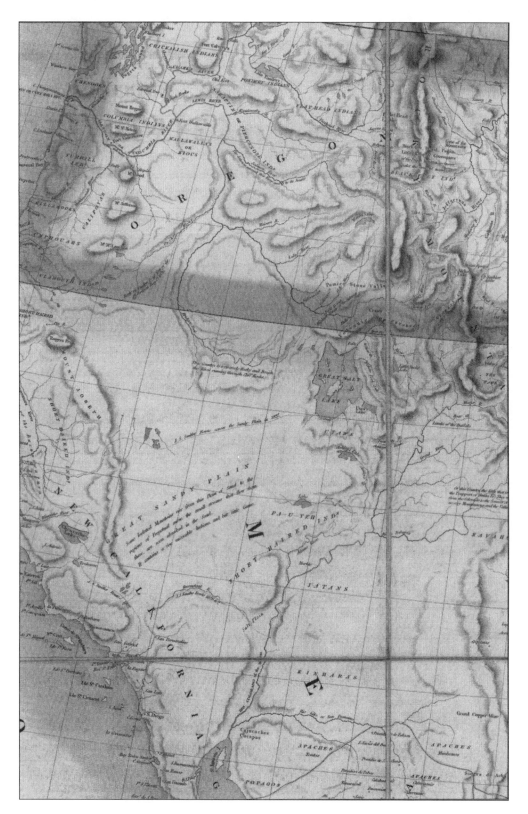

DETAIL OF FIGURE 5.3

originating in areas of "Perpetual Snow" are soon absorbed into the sands of the Interior West. Burr did not restrict his comments on this later map to the physical features. He noted, for example, that a "few miserable Indians" lived in this forlorn region.

In deciphering Burr's masterful map, we need to acknowledge the contributions of mountain men like the irrepressible Jedediah Smith, who probably knew the Interior West better than any other European American during this formative period of mapmaking. Historian Dale Morgan notes that Jedediah Smith had an important role in providing geographic information, as well as actual maps, that in turn influenced mainstream cartography. Smith traversed the Great Basin in 1827, his route leading through the heart of the region. According to Morgan, maps by Smith became part of a rapidly growing geographic knowledge of the late 1820s and early 1830s. Morgan searched for, but was unable to locate, maps drafted by Smith for the Hudson's Bay Company ca. 1828-29. He concluded that "Jedediah's final map appears to be irretrievably lost, known only in the form of David H. Burr's borrowings from it."[8] It should come as no surprise that Jedediah Smith's work was appropriated, as it were, by those recognized as geographers—that is, those who systematically accumulated, and then portrayed in map form, discoveries by those in the field. After all, Smith was an adventurer and seeker, not a formal presenter of such knowledge.

Burr's masterful map is thus a tribute not only to recognized cartographers, but also to those unacknowledged individuals who had a role in the creation of such landmark maps. Compared with other maps of this period, Burr's map is restrained, actually cautious. Its accuracy is attributable to Smith as its main source of information. According to cartographic historian Kenneth Nebenzahl, Jedediah Smith had provided surprisingly accurate information at a time when speculation about the West's geography was still rampant. As Nebenzahl notes, Smith would have entered the "pantheon of American pathfinders that include Lewis and Clark, Pike, Long, and Fremont [sic]" had he not been murdered by Comanches on the Santa Fe Trail just shy of his thirty-third birthday.[9] For his part, Burr fared better. He is remembered and revered by cartographic historians as a master mapmaker. Equally important was Burr's political stature: as the official geographer of the U.S. House of Representatives, his maps were accorded considerable authority.

Interestingly, numerous mapmakers stubbornly refused to alter their products to include geographic revelations like those seen on Burr's maps. Others, however, reveal tentative transitions away from the older, if not obsolete, maps. Consider, for example, the Society for the Diffusion of Useful Knowledge's 1842 map *Central America II, Including Texas, California, and the Northern States of Mexico* (fig. 5.4). Published four years *after* Burr's, it does indicate areas of interior drainage, including "Lake Youla Salt," presumably the Great Salt Lake. However, it also contains considerable mythical information, including the Rio Buenaventura flowing to the Pacific Ocean. In fact, it shows *two* rivers bearing the name Buenaventura. But note that the courses of both of these streams are now tentative in that they are indicated by *dashed* lines. The map shows yet another river—the Rio Pescador— running from the Interior West all the way to San Francisco Bay. This type of mythical geographic information on maps may seem downright wrong to us today, but it serves a purpose in that it is often deeply rooted in broader mythology. Take the notion of a large lake lying at a continent's very interior—the most difficult place to reach. It was not new at all. Deeply embedded in popular mythology, it lured explorers into the interiors of most continents. In South America, for example, Jan Jansson's map of Guiana (1635) places Parima Lake in the north-central part of the continent, a remote area of indefinite hydrology that was central to Sir Walter Raleigh's fabled El Dorado.[10] In this case, El Dorado is a material manifestation of a much earlier version—the Garden of Eden. In a metaphorical sense, it serves a similar purpose to a navel, a tangible point of connection to sources of origin. Given the information that found its way into the consciousness of explorers about the interior lake in the western part of North America, it is almost predictable that the remote interior here would also resonate as a point of human origins.

That is exactly what happened as explorers and colonizers sought to locate the stuff of legends, namely Aztlán—the Aztecs' toponym for their ancestral place of origin somewhere northwestward of Mexico City. The description of Aztlán, or "Place of the Heron Feathers," is as evocative as it is nebulous. It is associated with the color white, which in turn had spiritual significance to the Aztecs. Because Aztlán is "in a lake, reminiscent of the site of their great city to be, Tenochtitlán," certain locations in the Great Basin, as well as many other sites from California to central Mexico, were

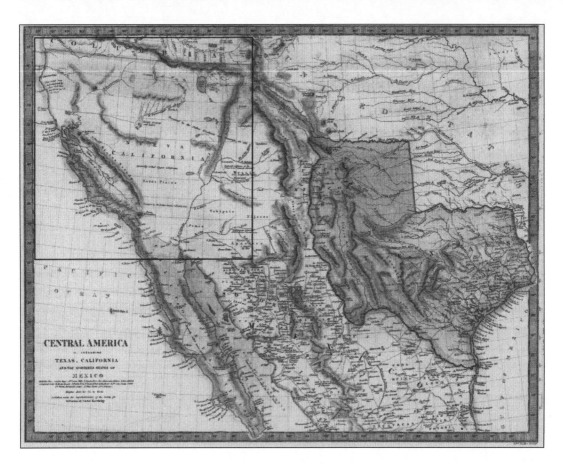

FIGURE 5.4.

Society for the Diffusion of Useful Knowledge, *Central America II,*
*Including Texas, California, and the Northern States of Mexico* (1842). Courtesy Special
Collections Division, University of Texas at Arlington Libraries

candidates. Many placed Aztlán not far from the densely settled pueblos of
New Mexico, but still others thought it lay farther north. Miera's comparison of the area around Utah Lake (with its abundant waterfowl) to the
Valley of Mexico helped to conflate Aztlán with this part of North America;
however, many scholars believe the most promising candidates are lakes in
the states of Michoacán and Jalisco, or perhaps in the state of Nyarit.[11] These
lakes are considerably farther south than the Great Basin, but in the late
eighteenth and early nineteenth century the cartographic imagination was
vivid indeed. The demands of science and observation were still, at times at
least, subservient to romance and speculation. As the North American interior began to be articulated on maps, likely with the help of Native
American informants, cartographers now tentatively placed Aztlán on their

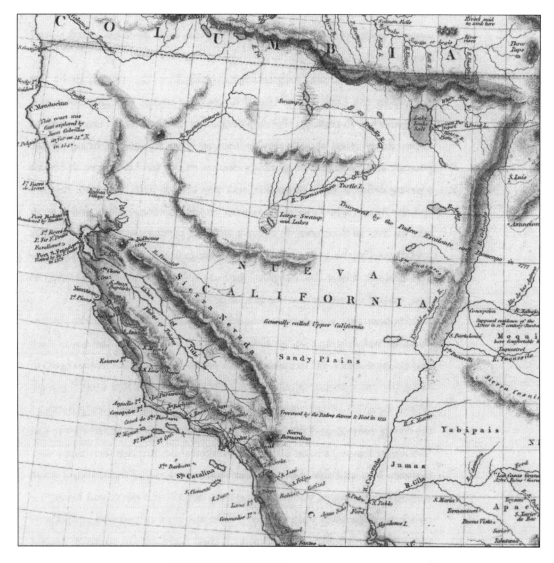

DETAIL OF FIGURE 5.4

maps. And why not? As we have seen, the Spaniards themselves had compared the area around Utah Lake with that of Tenochtitlán in the 1770s. From there, the legend took root. As noted earlier, Alexander von Humboldt's map states that an area east of the Utah Lake oasis is the supposed homeland of the Aztecs. The Society for the Diffusion of Useful Knowledge endorsed the "supposed residence of the Aztecs in 12th century (Humboldt)" on their 1842 map, thus further validating this concept in the popular imagination. There is something decidedly comforting about a mysterious source of origins deeply hidden in an unknown interior, for it suggests nothing less than the source of our own personal origins—the womb.

Decidedly retro and pregnant with symbolism, the 1842 map by the Society for the Diffusion of Useful Knowledge makes an important point. Our tendency to consider this map's depiction of nonexistent rivers as erroneous information should also be tempered by an understanding of the potency and durability of an image once seen but hard to forget. It reveals a strong desire of both mapmaker and map user to believe in both the past and the future. The fanciful depictions on maps become icons, and as such take on lives of their own. In other words, the persistence of graphic images is explained by their becoming *ideas* that are increasingly embedded in popular thought. Once this happens, they are regarded and then respected as true, even if the image is, in reality, incorrect.

It is here that we encounter another popular watershed in the understanding and mapping of the Great Basin—the coining of the name itself and its appearance on manuscript and printed maps of the mid-1840s. In 1844, the enterprising explorer John Charles Frémont first used the term "Great Basin" for "that anomalous feature in our continent" lying between the Sierra Nevada and the Wasatch Mountains. Frémont added that it was "a singular feature; a basin some five hundred miles [in] diameter every way, between four and five thousand feet above the level of the sea, shut in all around by mountains, with its own system of lakes and rivers, and having no connection whatever with the sea."[12] The way Frémont arrived at, and promoted, this conclusion is fascinating. It sheds light on the complex process of regional image-making. Frémont's knowledge was based both on his own field expeditions and on information provided by Joseph Reddeford Walker, whom Frémont credits in his reports and memoirs as first to recognize the region's interior drainage. Frémont was indeed what his popular name "Pathfinder" suggested—someone who helped lead the national psyche into understanding the premier geographic secret of the western frontier. Frémont deserves credit for popularizing the Great Basin following several strenuous expeditions around and through it—this despite the fact that many historians consider his character questionable and his geographic accomplishments minimal.

But if Frémont became obsessed with deciphering the Intermountain West, he had considerable help in doing so from often overlooked sources.

As Frémont aggressively sought to understand the lay of the land, and thereby ultimately to decipher the region, he had to evaluate a growing amount of geographic information that accumulated with each report he read, each informant he encountered, and each mile he traveled. This information gathering, it should be noted, involved a fertile cross-cultural exchange with Native Americans. On December 6, 1843, Frémont recorded how he learned about the course of the Pit River southeast of Klamath Lake in today's Oregon. Confused by the river's turning from easterly to south, Frémont's expedition was assisted by the Klamath Indians. In his own words, "drawing a course upon the ground, they made us comprehend that it pursued its way for a long distance in that direction, uniting with many other streams, and gradually becoming a great river. Without the subsequent information, which confirmed the opinion, we became immediately satisfied that this water formed the principal stream of the Sacramento River."[13] Frémont realized that he was at the northwestern edge of the Great Basin, for the Pit River flowed westward to the Pacific, while the waters just east of this point flowed eastward into the interior. The Indians had helped him solve a problem here, but Frémont was not always as forthcoming with praise as he might have been. On January 15, 1844, he asked the Northern Paiute for "information respecting the country," observing that "we could obtain from them but little." And yet Frémont quickly adds, almost matter-of-factly, that the Indians "made on the ground a drawing of the [Truckee] river which they represented as issuing from another lake [Tahoe] in the mountains three or four days distant, in a direction west of south; beyond which, they drew a mountain [Sierra Nevada]; and further still, two rivers [Sacramento and/or American or Feather or San Joaquin] on one of which [Sacramento?] they told us that people like ourselves traveled."[14]

Frémont was obsessed with what lay in the interior of the region, but much of it remained blank on the map originally published with his 1845 report. With each subsequent expedition, more was learned and added to maps. Fortunately, Frémont wisely selected one of the finest cartographers for the task of mapping his expedition's findings—George Karl Ludwig Preuss. Born in Höhscheid, Germany, in 1803, Preuss was called Charles and took his mapping duties seriously. Charles Preuss's map from Frémont's expedition (fig. 5.5) reveals the complex process of mapmaking in the 1840s and was one of the most influential to evolve during this period. It now

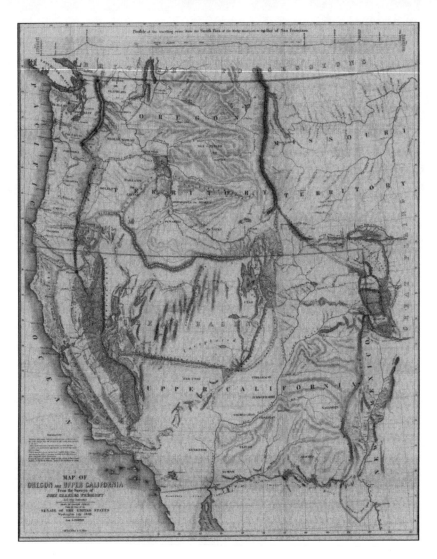

FIGURE 5.5.

Charles Preuss, *Map of Oregon and Upper California from the*
*Surveys of John Charles Frémont and Other Authorities* (1848).
Courtesy David Rumsey Collection

depicts visually, and describes in words, the region as "filled with rivers and
lakes which have no communication with the sea." That statement confirmed what many had begun to suspect and Frémont publicly verified—
that this is a land of interior drainage whose waters never reach the ocean.
Frémont's name "Great Basin" was not only ingenious, it has endured as
shorthand for the solution to a centuries-old geographic mystery.

Frémont also deserves credit for something equally important that his

detractors failed to note. At the same time that Frémont-Preuss maps brand-
ed the region as the Great Basin, Frémont also published several lithographs
of geographic features that artists on his expedition, including Preuss, had
drawn. These features became landmarks to traveler and map reader alike.
Among these landmarks was Pyramid Lake, with its evocatively shaped Tufa
Island that so stirred the imagination (fig. 5.6). Frémont's illustrations combined with his map to increase visual expectations of those exploring the
Intermountain West. From that time onward, the public would no longer be
satisfied with only a map, or only verbal description; rather, it would
demand, and receive, illustrations of features that so excited the imagination.
Like maps, these images would be recycled and abstracted. Thus, Pyramid
Lake in Emmanuel Henri Dieudonné Domenech's *Voyage Pittoresque dans les
Grands Deserts du Nouveau Monde* (fig. 5.7) is striking—apparently derived
from Frémont's narratives, but even more fanciful; so, too, was Domenech's
narrative description of the Great Basin itself influenced by the words of the
Pathfinder, but rendered even more impressive for his European audience.
Domenech typified the international interest in the Interior West. Born in
Lyon, France, in 1825, he was ordained a priest in San Antonio, Texas, in 1848
and became intensely interested in the American West. Domenech's fascinating book *Seven Years' Residence in the Great Deserts [of North America]* was
printed in both English and French and became popular on two continents.

At the time that Frémont was exploring the Great Basin, a new form of
recording places—photography—was just developing. Historian Ron Tyler
writes that Frémont tried this new medium but failed miserably in recording images. The daguerreotypes that Frémont attempted to produce in the
field as early as 1842 left him disappointed, likely because the difficulties of
controlling conditions properly were overwhelming. Frémont's experience
left him disillusioned—so much so that he and many subsequent Western
explorers did not use photography; instead, they relied on the tried and true
methods of pencil and pen on paper, with an occasional watercolor when
the scenery or some other object of interest demanded it.[15] An indication of
the tensions between Frémont and Preuss surfaced early on about this very
issue of recording the landscape on daguerreotype plates. Preuss's reaction is
noteworthy, for it says much about his temperament. Commenting on how
Frémont had "spoiled five plates," Preuss concluded, "That's the way it often
is with these Americans. They know everything, they can do everything,

FIGURE 5.6.

Pyramid Lake, from Frémont's Report (1844).
Courtesy Special Collections Division, University of Texas at Arlington Libraries

and when they are put to the test, they fail miserably."[16] On this first expedition into the Great Basin, tempers flared frequently. Preuss was temperamental enough, but Frémont was also mercurial and developing a long list
of detractors as he traveled in search of answers about the Great Basin's mysterious geography.

But no name in the nineteenth century better typifies the popularization
of the Great Basin than Frémont. As a leader of several daring expeditions
to the region, he cut a romantic figure. Frémont is responsible for naming,
and hence claiming, many of the region's features. Symbolically, the Buenaventura River now becomes the Humboldt River on Frémont's map. This is
Frémont's tribute to Alexander von Humboldt as scientist and geographer.
Even though Humboldt's own maps of 1804-1812 do not accurately depict
the region's hydrology, Frémont remedies this as the river now bears
Humboldt's name. Frémont also, in effect, aggrandizes his own accomplishments by associating them with the recognized genius Humboldt. Some
might say it was inevitable that Frémont would name a river and a mountain range after his brilliant German mentor, for by so doing, he immortalized Humboldt and all who respected him.

FIGURE 5.7.

Pyramid Lake in E. Henry Domenech's *Voyage Pittoresque dans les Grands Deserts du Nouveau Monde* (1862). Courtesy Special Collections Division, University of Texas at Arlington Libraries

And yet, despite Frémont's seeming obsession with Humboldtian science, some of his own maps—or rather the Frémont-Preuss maps—also contain erroneous information. One of the most fascinating errors is a prominent transverse mountain range running across the Great Basin region, when in fact the mountains here trend north-south. This mountain range on the Frémont-Preuss map is not only unique, but somewhat bizarre, in that it seems to counter all observation. A notation on the 1848 map suggests its tentativeness, for the mountain range was only seen from a distance and not explored. Drawn as bold feature stretching more than two hundred miles in an east-west direction on the map, it runs cross-grain to the actual topography. According to a historian of the Great Basin, Frémont's east-west mountain range was "apocryphal."[17] That is a good term, for it has two separate but related meanings. The first, something of doubtful or dubious authenticity, is secular in nature. The second, however, refers to religious writings that were excluded from the Bible. This latter definition suggests *belief* on the part of those who drew the mountain range without ever having experienced it. In other words, although the apocryphal feature was later considered dubious, it was first drawn on faith.

Where did this mountain range originate? Had the Frémont party heard about it from the Indians or mountain men? Did they accidentally reorient one of the north-south-trending ranges? A hint of a real transverse mountain range within the Great Basin is found in an area traversed by Frémont. At the western edge of the Mojave Desert near Walker Pass, the El Paso Mountains obediently line up along the Garlock Fault. That active fault can be traced eastward for more than one hundred miles, and along it several smaller ranges line up to counter the generally northwest-southeast trend of the topography. Interestingly, but perhaps coincidentally, Frémont Peak overlooks this peculiar geologic anomaly. From this peak, one gazes north to see the El Paso Mountains lying almost perfectly east-west. Looking in either direction, one can sense the awesome geological forces that align the ranges longitudinally here while most everything else in the region is oriented latitudinally. Geologists speculate that this longitudinal rift is traceable to the Cretaceous (ca. 65 million years ago) and that it has controlled subsequent episodes of mountain building.[18] We can only speculate whether this anomaly was the source of Frémont's transverse range; however, its general orientation and rough geographical position in the southern Great Basin suggests that it might have influenced his imagination and his map-making. It is possible that Frémont's expedition was so impressed by the transverse ranges at the south edge of the Mojave Desert that they rendered one into a cartographic icon. If so, however, they placed it about 150 miles too far to the north on their maps.

But Frémont's mysterious transverse-running mountain range may have yet another origin. Hints of it are seen on some earlier maps of the region, and cartographers associated with Frémont's expeditions may have emulated those errors. Frémont himself seems to caution map readers, for he admits that he did not actually explore the range, but only saw it from a distance. This admission, perhaps, honestly suggests a good deal of conjecture about both the feature's location and its form, but once put on paper an image is hard to forget. The Frémont map's text indicates that this feature is a "dividing range" between the waters that sink into the region's sands and those that find their way southward to the Pacific via the Colorado River. It is possible that explorer Frémont and cartographer Preuss used the term *range* to indicate a land area oriented in a certain direction, rather than literally a formidable mountain range like those that elsewhere bound the region (notably, the Sierra Nevada and Wasatch ranges). In other words, *range* can here mean something different from its commonly understood definition (a series of mountains), but rather more its initial meaning—a series of things oriented in a relatively straight line, or even an area oriented in a certain direction, as well as an area that can be ranged over. In any event, Frémont knew that a *watershed* was reached toward the southern portion of the Great Basin, that here lay a dividing line, hydrologically speaking. Words and images would have very likely perpetuated this critical divide as a mountain range in the vivid geographic imagination of the mid-1840s.

Regardless of their inevitable errors, Frémont's expedition and maps represented a defining moment in the region's cartographic history. By the mid-1840s, intrepid travelers seeking homes in California were crossing the central portion of the Great Basin as part of the now legendary "Overland Trail." In 1846, when Samuel Augustus Mitchell published *A New Map of Texas, Oregon, and California,* he definitively called the region the "Great Interior Basin of California." Mitchell's map concisely states that "streams and rivers here have no outlet to the sea."

Despite the region's formidable character, it attracted the Mormons, who hoped to find a place beyond the Rocky Mountains to practice their religion unmolested. In preparing the Latter-day Saints for the move westward, Brigham Young consulted the most recent official reports, emigrant guides, and maps that he could find. Writing to Joseph Stratton of St. Louis in February of 1847, Young was clear about the type of maps he wanted: "Bring me one half dozen of Mitchell's new map of Texas, Oregon, and California and the regions adjoining, or his accompaniment for the same for 1846, or rather the latest edition and best map of all the Indian countries in North America." Anxious to get the most accurate information, and not sentimental about the products of any particular mapmaker, Young added: "If there is anything later or better than Mitchell's, I want the best."[19] In anticipation of the move to the Intermountain West, Young anticipated one cartographic product in particular—Frémont's map and the report that it accompanied. As he led the Mormons in settling vast areas of the Intermountain West, Young thus continued to rely on maps produced by a government he increasingly distrusted.

Mitchell's map rendered the Great Basin as one of the Interior West's landmark features. It and the Frémont-Preuss map were widely consulted,

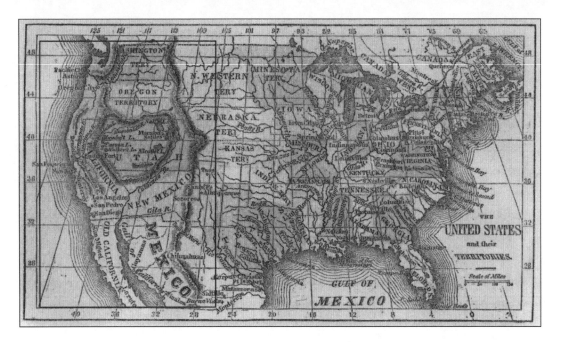

FIGURE 5.8.

Marcius Willson, *The United States and Their Territories* (1854).
Courtesy Nancy Grace, Weatherford, Texas

setting the standard. John Disturnell's influential 1847 *Mapa de los Estados Unidos de Mejico* followed suit, confirming that the Great Basin had now become an icon. Interestingly, by 1850, the U.S. Bureau of the Corps of Topographic Engineers' map showing the region reveals glimpses of a cartographic change under way: the transverse mountain range that was shown so boldly on the Frémont-Preuss map is now shown entirely using only a dotted line—a prelude to this mythical feature's impending disappearance from maps.

The process by which one map affects others is fascinating and revealing. When the Frémont-Preuss map was first published, it created a demand. In the space of several years, the information it contained was added to numerous government maps. The Frémont-Preuss map also rendered earlier commercial maps obsolete. But because mapmakers had considerable capital invested in their plates, they often altered only the portion requiring revision. On its 1846 edition of *Central America II,* for example, the Society for the Diffusion of Useful Knowledge indicated that "the unexplored region bounded on the W[est] by the Sierra Nevada and on the E[ast] by the Bear River and Wahsatch Mts. has been called the great interior basin of

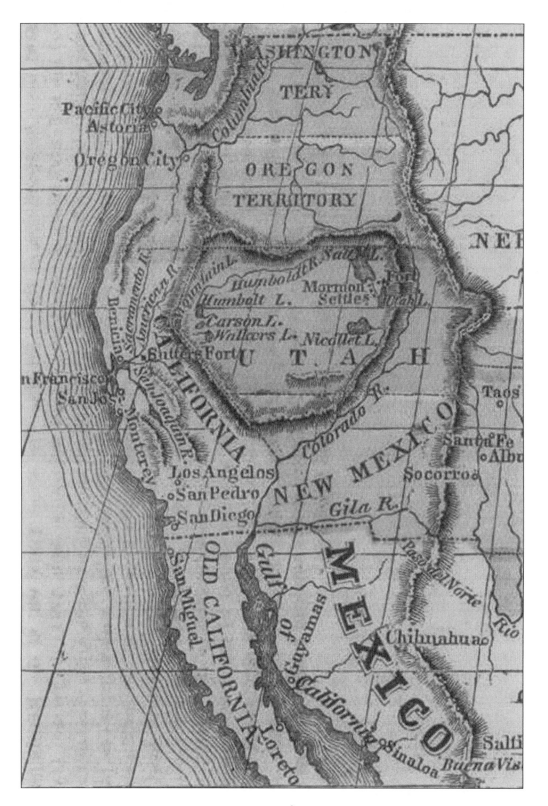

DETAIL OF FIGURE 5.8

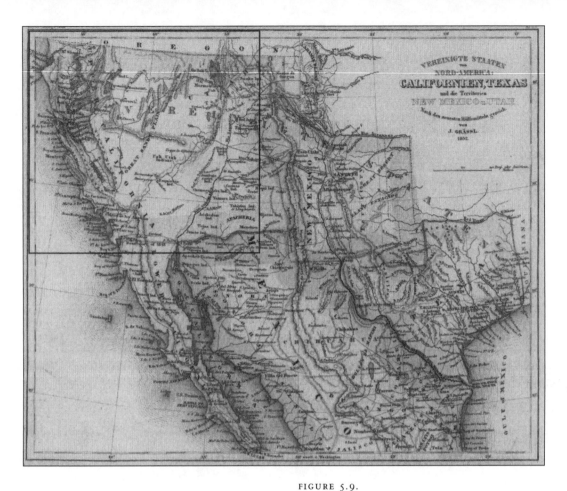

FIGURE 5.9.

*Vereinigte Staaten von Nord-America: Californien, Texas, und die Territorien* (1852).
Courtesy Special Collections Division, University of Texas at Arlington Libraries

California."[20] The publisher, Chapman and Hall of London, seamlessly integrated the new information, while the rest of the map remained virtually unchanged. This is not only a tribute to Frémont but a nod to both accuracy and marketing. The public was becoming both more discerning and more demanding.

Pressures on exploring expeditions at this time were immense, for they now were subject to two demands—that of spectacular revelations about claimed lands, and sobering accuracy in what they related. The Frémont-Preuss partnership is fascinating and reveals differences in temperament between explorer and cartographer. Whereas the ebullient Frémont thrived in the field and seemed to be energized by each challenge, the moody Preuss found setbacks disconcerting and longed for the comforts of home. He

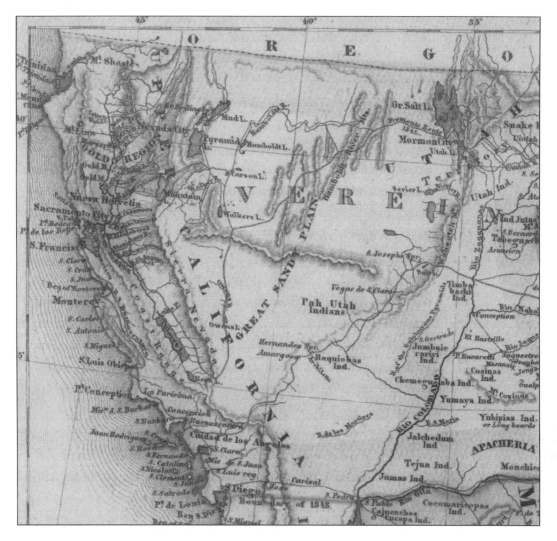

DETAIL OF FIGURE 5.9

complained almost continually about Frémont, but that should not detract from Preuss's cartographic accomplishments. He helped an anxious America visualize the geography in the Interior West in the mid-to-late 1840s and early 1850s. In addition to being a fine cartographer, Preuss was a superb artist-illustrator. Several of his landscape sketches accompanied Frémont's report, and they reveal the nearly perfect coordination of scientific mind and artistic hand. Preuss's *Map of Oregon and Upper California from the Surveys of John Charles Frémont and Other Authorities* (1848) is widely regarded as one of the most significant cartographic accomplishments in the history of the West. Preuss was recognized in his time as a master mapmaker. However, by

1854, even recognition and the comforts of home could not stabilize Frémont's tormented cartographer, who hanged himself from a tree to end his despair.[21]

By the mid-1850s, the map-reading public was becoming more discerning, but printed maps produced by commercial publishers often lagged behind the increasingly accurate maps included in original governmental reports. That lapse in part explains some cartographic discrepancies and curiosities. Some popular maps took Frémont's name "Great Basin" to mean exactly that. For example, on the map *The United States and Their Territories* (fig. 5.8) that appeared in Marcius Willson's 1854 *Outlines of History: Illustrated by Numerous Geographical and Historical Notes and Maps,* the Great Basin is depicted as a virtually empty bowl completely surrounded by tall mountains. Although it does contain a hint of the transverse mountain range, overall one is impressed by just how much of a single depressed region the Great Basin is on this and other maps like it. And why not? Frémont's phrase "Great Basin" is so evocative, and so suggestive of a single large interior-drained basin or sink, that impressionable cartographers naturally were tempted to render his words into image.

Willson was a prolific producer of books on North American history that contained informative maps. His map deserves a closer look because it reveals so much about the way words can shape images. The only topography marring the otherwise perfect basin on this map is that range of transversely aligned mountains in the southern part of the region—the legendary Dividing Range promulgated by Frémont himself. This range would appear on maps well into the 1850s and would influence not only Americans but Europeans as well. Consider, for example, a German map showing the *Vereinigte Staaten von Nord-America: Californien, Texas, und die Territorien* (United States of North America: California, Texas, and the Territories).[22] Published in 1852 and subsequent years until the mid-1850s, this map features a Great Basin adorned with the mysterious, but nonexistent, transverse-trending mountain range (fig. 5.9). These deviations from the actual geography fascinate us today, but they should be seen as essential steps in the cartographic evolution of a popular region. Only through additional exploration and mapping would such errors disappear as increasingly accurate depiction of the region's hydrology helped demystify the topography.

If, by the mid-1850s, most maps recognized the Humboldt River as end-ing in the western portion of the Great Basin and the Dividing Range became but a memory, this increasing accuracy coincided with the demands of an increasingly scientific mindset. As myth yielded to science, and conjecture to progress, those increasing expectations of accuracy would soon brand all maps of the Great Basin. The reasons were at once scientific and anthropological. This region was not only home to Native Americans whose presence was becoming problematical as the frontier expanded into the Intermountain West, but it had also become the site of a new and unique city religious community—Salt Lake City—that now appeared on maps at the eastern edge of the bowl-shaped Great Basin. That city and its occupants, the Mormons (as members of the Church of Jesus Christ of Latter-day Saints are called), fascinated Americans and Europeans alike. One can see the Mormon presence in the West take shape on popular maps. Whereas Marius Willson's popular history books and their accompanying maps in the mid-to-late 1840s reveal the Great Basin to be a desolate, nearly blank space, his 1854 map shows the presence of "Mormon Settl[ement]s"—each marked with a black **X**. Placing the followers of a new, mysterious American religion on maps along with other peculiar features—like rivers that never reached the sea, and lakes even saltier than the ocean—guaranteed that the Great Basin would continue to hold the interest of text and map readers for another generation to come.

# 6

## Maps in the Sand
### 1850–1865

EXPLORERS HELPED DEMYSTIFY the Great Basin by the mid-1840s, but the pace of demystification dramatically increased as westward-moving pioneers traversed the region shortly thereafter. News and stories about the pioneers' exploits were eagerly awaited by those who remained in the East as the drama of western settlement took center stage. Nothing helps to increase knowledge of places more than the firsthand experience of colonists and travelers, and those travelers filled pages of text and drew new lines on well-worn maps such as the Frémont-Preuss map of 1848. Two very different kinds of settlers experienced the Great Basin at this time—the Mormons, who proclaimed "This is the place" of their new home in 1847, and would-be miners flocking to the California goldfields, who trekked across the region in 1849. Both groups did much to call attention to the Great Basin.

At the time of the California gold rush, the southern portion of the Great Basin was traversed by travelers moving along the Old Spanish Trail from San Bernardino eastward to Las Vegas and beyond, but large areas still remained unexplored. This section attained some notoriety when a California-bound party of gold rush immigrants elected to take a diagonal course across the region and enter California's Central Valley via the snow-free Walker Pass. This party's misfortunes became part of the folklore of the Intermountain West, but the role played by geographic information—or rather misinformation—in contributing to their dilemma has not been adequately discussed. Although the expedition's chronicler William Lewis Manly recorded his recollections of the event in 1894—more than forty years after the fact—they vividly reveal what can go wrong when people place too much faith in maps.

In his classic memoir, *Death Valley in '49,* Manly provided a cautionary reminder for his late-nineteenth-century public: "Reading [that is, literate] people of to-day, who know so well the geography of the American continent, may need to stop and think that in 1849 the whole region west of the Missouri River was very little known." Manly explained that friendly Indians' knowledge of the countryside helped avert disaster as his party reached the Intermountain West. He noted that one Indian in particular— Chief Walker—had a remarkable understanding of the geography of parts of the region they were about to traverse: "I have often wondered at the knowledge of this man respecting the country, of which he was able to make us a good map in the sand, point out to us the impassible cañon, locate the hostile Indians, and many points which were not accurately known by our own explorers for many years afterward."[1]

As Manly's group traveled toward the southern edge of the Great Basin, they were overtaken by another wagon train under the command of a Captain Smith. This party "had a map with them made by one Williams of Salt Lake[,] a mountaineer who was represented to know all the routes through all the mountains of Utah." According to Manly, this map "showed a way to turn off the southern route not far from the divide which separated the waters of the [Great] basin from those which flowed toward the Colorado, and pass over the mountains [Sierra Nevada], coming out in what they called Tulare Valley." As the travelers pondered their choices, Smith's map "was quite frequently exhibited and the matter freely discussed in camp, indeed speeches were made in the interest of the cut-off route which was to be much shorter." The map proved to be seductive, for it seemed so accurate. Manly recalled that it "showed every camp on the road and showed where there was water and grass, and as to obstacles to the wagons it was thought they could easily be overcome." Not everyone agreed— "captain Hunt felt that the route on Smith's map was not safe"—but Manly's party took it, and found "the country to grow more barren as we progressed."[2]

Having seen that map, an imaginary geography impressed itself into the minds and hopes of those who followed its guidance. Enduring many hardships, the party traversed numerous mountain ranges and found themselves disoriented in the western Great Basin. Here they spied a range they thought was the Sierra Nevada and acted on that belief. Having read Frémont's accounts of the fertile lands just over this range, they pressed on. But alas, when they crossed the mountains, they found themselves in yet another valley, this one even more formidable than the others. To commemorate the losses that the group suffered, they called it Death Valley—a name so dramatic that it was destined to endure. Manly and the rest evidently realized too late that the Williams map was not only useless, but downright treacherous. Looking back over the landscape they had crossed, one of the group declared, "Just look at the cursed country we have come over!" Manly characterized their dilemma of facing "a desert stretching out before us like a small sea, with no hope of relief except at the end of a struggle that seemed hopeless . . . more than any pen can paint, or at all describe. . . . [L]et your own imagination do the rest."[3]

The expedition's tribulations called for more than words. As if to set the record straight, their harrowing ordeal produced a map—albeit long after the expedition crossed the Great Basin's "grand, but worthless landscape" and safely reached California's bucolic Central Valley.[4] This map (fig. 6.1) was primitive by the changing standards of the late nineteenth century, but the fact that it was compiled by a survivor of a harrowing ordeal gave it credibility. The map's evocative title—*Showing the Trail the Emigrants Travailed* [sic] *from Salt Lake to San Bernardino in 1849*[5]—suggests that it will not merely position geographic features, but also demarcate a number of places and events linked to extreme hardship. These include "Mount Misery here the first death occured," and places where others died. As if to add an even more ominous touch, Mountain Meadows—the site of the infamous 1857 massacre of California-bound Missourians by Mormons and Indians—is also shown. The odd usage in the map's title—"travailed" where we might expect "traveled"—is more than coincidental. The word *travel* in fact derives from *travail,* a reminder how arduous traveling was in centuries past. The map itself reminds us that maps can also be compiled after the fact to commemorate some event of significance, in this case a disastrous trek across the Great Basin involving great hardship. That terrible journey had already given the region a reputation as punishing country. Nevertheless, its vast

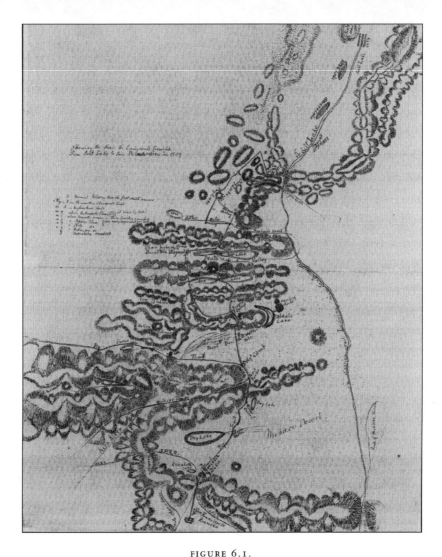

FIGURE 6.1.

Map *Showing the Trail the Emigrants Travailed [sic] from Salt Lake to San Bernardino in 1849*. Courtesy The Huntington Library, San Marino, CA

size and formidable geography taxed Anglo-Americans' abilities to comprehend it all.

It is one thing to draw sweeping, oversimplified maps of a region, but quite another to comprehend the intricacies and subtleties of its geography/topography. Traversing the central part of the region about the same time as Manly's ill-fated journey (1849), J. Goldsborough Bruff described another map drawn in the sand, as it were, by a Native American informant. According to Bruff, the map was created by "an aged Indian" who had visited the camp and seemed to know a lot about the country. Hopeful that the Indian could shed some light on the subject of their search, Bruff's party asked him about "a deep-basined lake where there was gold." To leave little doubt about what they meant, they "showed him a lump of the metal" that they sought. The "old savage" at this point "took a pair of macheres (large flat leathers to throw over the saddle) and sprinkled sand over them, drew a model map of the country there, and beyond it, some distance." In doing so, Bruff notes, the Indian "heaped up sand, to form buttes, and ranges of mountains; and with a straw, drew streams, lakes and trails: then adjusted it to correspond with the cardinal points, and explained it." These descriptions of Native American maps of the area reveal that the first raised relief maps of the Great Basin—that is, maps that literally vary their surface to depict topography—were made by the indigenous peoples.

Consider the wealth of information that this Indian provided. According to Bruff:

> He pointed to the sun, and by signs made them understand, the number
> of day's travel from one point to another. On it he had traced, (as I found
> on their explanation,) Mary's Humboldt River, Carson River, Pyramid
> lake, and the emigrant routes,—above and below. He moved his finger,
> explanatory of the revolutions of wagon wheels, and that white people
> travelled along, with guns, on the said routes. On his map, he had
> exhibited the lake they were then at, and another in a deep basin, with
> 3 buttes beside it, and said that gold was plentiful there; and also that
> 10 months ago the whites had visited it, and fought with the Indians.[6]

Bruff was no ordinary traveler. His stunning sketches of the Great Basin (fig. 6.2) reveal his awareness of the topography and geology—and again demonstrate that illustration and mapmaking are close companions. Bruff was, in fact, a cartographer of some note: He had drafted numerous maps during the U.S.-Mexican War (1846-48)[7] and was adept at depicting the lay of the land in rugged western areas. His description of the old Indian's map is fascinating, for it is a tribute from one mapmaker to another. The fact that the two mapmakers were from radically different cultures makes Bruff's tribute all the more remarkable. His account of the old Indian's mapmaking skills confirms what many seasoned travelers knew: the Native Americans understood the region more intimately than did the most knowledgeable of Anglo-American explorers.

FIGURE 6.2.

J. Goldsborough Bruff, sketch of
"Great Basin—Mountains on the Humboldt (Mary's) River" (September 10, 1849).
Courtesy The Huntington Library, San Marino, CA

Mormon exploration parties also found the Native Americans' geographic knowledge helpful. As the Parley P. Pratt exploring party worked its way southwestward from Salt Lake City in 1849-50, it relied on a sketch map that the group's surveyor-engineer, W. W. Phelps, had made after climbing to the summit of Mount Nebo. The Pratt party was especially interested in the country that lay to the southwest near Sevier Lake. When Phelps's map was shown to Indian leader Chief Walker (Wakara), those present were "all astonished" at his understanding of the geography. As the expedition's clerk noted, Walker "showd [sic] points in it & told [us] what country he was acquainted & what he was not, like an experienced geographer."[8]

Although the Great Basin had yet to be thoroughly explored, at least by Anglo-Americans, that condition was changing rapidly in the 1850s. A relatively new form of transportation—the railroad—was about to transform geographic knowledge of the region. This was the age of the iron horse,

the Humboldt (Mary's) River

whose iron rails would soon bind the nation to the Pacific Coast. Five rail-road routes were mandated by an act of Congress in 1852 and surveyed in 1853-54. Two of them—the routes along the 38-39th parallel and along the 41st-42nd parallel—crossed the Great Basin. These surveys built on earlier reconnaissances for wagon routes in the late 1840s and early 1850s.

In 1852, Captain Howard Stansbury produced the masterful *Map of the Great Salt Lake and Adjacent Country in the Territory of Utah,* which has been called "the first accurate and detailed geographical account of the Wasatch Oasis, a region destined to become the center of Mormon culture region and the major metropolitan area of the Great Basin."[9] Stansbury's emphasis on the region's unique geology and hydrology was groundbreaking; his report entitled *Exploration and Survey of the Valley of the Great Salt Lake of Utah, Including a Reconnaissance of a New Route through the Rocky Mountains* has become a classic in the region's literature. Stansbury's expedition brought an end to speculation. Beginning in October of 1849, Stansbury's expedition was the first to successfully circumnavigate the entire lake and determine its true character. The expedition involved teamwork: Stansbury focused on the Great Salt Lake while Lieutenant John W. Gunnison covered the area to the south—a decision that proved to be fateful indeed.

Surveying the Great Basin at this time became extremely dangerous as tensions between Mormons, non-Mormons, and Indians mounted. Lieutenant Gunnison's name would soon be associated with exploration's successes and tragedies. A graduate of West Point in 1837, Gunnison became a well-respected surveyor on the Stansbury expedition, which he joined in 1849. In 1853, his role shifted to leader of the survey team searching for a railroad route to the Pacific. Taking an interest in all aspects of the region, Gunnison had written a treatise on the Mormons that seemed neutral enough to him but reportedly did not please Brigham Young. The Mormons were becoming hypersensitive about criticisms of them leveled by the "Gentiles" who now explored Deseret. On October 26, 1853, at a site near where the Sevier River reaches normally dry Sevier Lake, Gunnison's party was attacked by a band of Pahvant Indians. Within minutes, Gunnison and seven other members of the survey expedition were killed and their bodies mutilated. The attack reportedly occurred *after* the Gunnison party had been assured by Brigham Young and the Mormons that they were safe from Indian depredations. This brutal event only added to tensions between the Mormons and federal authorities. Today the desolate, forlorn site of the Gunnison Massacre is a sobering reminder of the perils faced in mapping the Great Basin (fig. 6.3).

Despite such setbacks, surveying of the region for railroad routes was irrepressible. Of the two routes across the Great Basin, the northerly one was projected to cross from Fort Bridger to the Pacific near San Francisco. Entering the region via either Weber Canyon or Timpanogos Canyon, this northern route ran past the Great Salt Lake to the Humboldt River, and then across the Sierra Nevada. The exploration and surveyors' report noted that "the country consists alternately of mountains, in more or less isolated ridges, and open level plains." Those mountains were formidable, but surveyors found passes through which the railroad could be built without excessive grades. The valleys would be easier to build a rail line through, but they were desolate, "merely sprinkled by several varieties of somber artemisia (wild sage), presenting the aspect of a dreary waste." The Humboldt River's course would enable a portion of the line to avoid steep grades. "The topographical features of the Great Basin," the report opti-

FIGURE 6.3.

Forlorn and vandalized marker commemorating the site of the Gunnison Massacre
west of Delta, Utah. April 2003 photo by author

mistically concluded, "present extraordinary facilities for the construction
of a railroad across it."[10] Surveyors immediately recognized that the second,
more southerly route, would present great difficulties in entering and crossing the Great Basin, and so they quickly ruled it out. They were soon proven
correct. The recommended route along the Humboldt was nearly ideal, and
the Central Pacific would complete its line through the region using that
route in 1869.

The exploration and survey report was accompanied by furious mapping
of the region. Some of the new cartographic products were merely base
maps upon which were superimposed geological and meteorological information. Produced at a scale of 1:7,450,000, the report's base map showed
only the crudest outline of a practicable and proposed railroad line. The
preferred route appears clearly on this map. As if oblivious to topography,
the line cuts directly through several north-south-trending mountain ranges
before following the easier Humboldt River course westward.[11] This, of
course, is an oversimplification. The map also contains some speculative
information, including that prominent east-west mountain range that the
Frémont-Preuss map first showed a decade earlier.

Clearly, more-detailed maps were needed to show Congress and others
just where the railroad would run. Those detailed maps, not coincidentally,
would also confirm that an increasing knowledge is a prerequisite for own-

FIGURE 6.4.

W. H. Emory, *Map of the Territory of the United States from the Mississippi to the Pacific Ocean* (1858). Courtesy Special Collections Division, University of Texas at Arlington Libraries

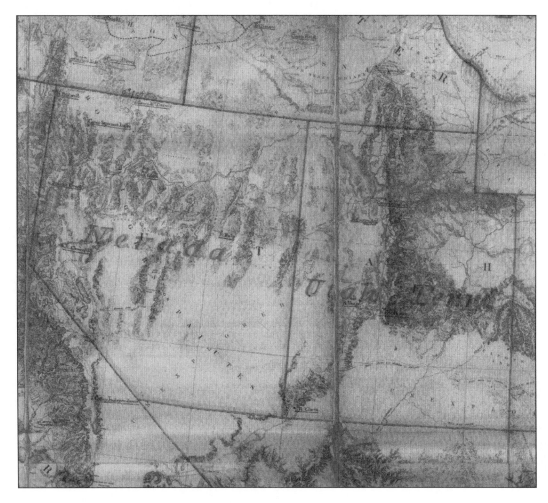

DETAIL OF FIGURE 6.4

ership. As secretary of war, Jefferson Davis ordered the preparation of such a map. He called it *Map of the Territory of the United States from the Mississippi to the Pacific Ocean to Accompany the Reports of the Explorations for a Railroad Route* (1857–58), and its preparation involved several of the best topographical engineers of the day. On this map, the area adjacent to the Humboldt River is highly detailed, but the southern portion of the region remained a mystery. There, amidst a huge blank space occupying most of southern Nevada, two phrases cross diagonally to form an **X**. These crossed phrases are startlingly reminiscent of those on Dufor's map a generation earlier. One phrase, running from southwest to northeast, indicates the area's status: "UNEXPLORED." The other phrase, which runs from northwest to southeast, is a tribute to the first peoples here: Boldly marked on the map, "PAIUTES" and then the fuller name "ROOT DIGGERS OR PAIUTES" stretch across the entire region. This term *diggers* we interpret today as a somewhat derogatory name for the native peoples, but the Anglo-Americans marveled that any people could actually subsist here—and the word does reveal their level of subsistence. Even the name Paiute is not the name that the natives called themselves. Rather, it is a white term for Ute Indians who knew the source or location of water (*pah*). In fact, the word *Paiutes* here acknowledges in effect not only that the area was unexplored by whites, but that it was well known to Indians. This map, by Major W. H. Emory, was updated by Captain A. A. Humphreys in 1857-58 and represented the state of geographic knowledge on the eve of Pacific railroad development.[12] This later map (fig. 6.4) shows large areas in the southern and northern Great Basin merely as "Unexplored Territory." Significantly, this wording replaces the names of native tribes at just the time the Indians were being marginalized.

This label "unexplored territory" commonly appears on maps of this period and demands closer inspection. The word *unexplored* here refers less to a permanent condition than to a condition needing rectification. At face value, *unexplored* means "unknown," but its placement on a map of exploration connotes "as yet unexplored." The word *territory* is also conditional. Although at first reading it simply suggests land, it subliminally identifies land in a certain condition or status—land that we subconsciously know will someday become part of a large political whole. Through both word and image, the "territory" on the map became part of Utah Territory. This seamless process of appropriation reminds us that "maps can be read as ideological assertions rather than as exact geographical representations."[13]

This map is part of a series of sectional maps. The sections of the route along the 41st parallel were drawn at an even larger scale—one inch to twelve miles—and their detail is impressive. On one sheet, topographer F. W. Egloffstein (his initials are incorrectly given as E. W. on the map) delineated the Great Salt Lake and the stark mountains west of it, which rise from an area labeled, with eloquent simplicity and candor, THE DESERT (fig. 6.5).[14] As in other maps, large areas away from the survey are blank, but this one designates the area to the south as "GOSHOOT INDIANS"—the subliminal message of white-on-Indian violence notwithstanding, a recognition of that Native American tribe's dominance here even as many other maps effectively eliminated them by erasure. Map number 3 covers the area

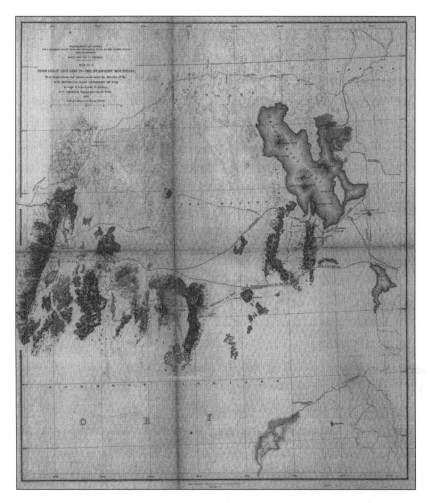

FIGURE 6.5.

E. [sic] W. Egloffstein, *From Great Salt Lake to the Humboldt Mountains; from Explorations and Surveys Made under the Direction of the Hon. Jefferson Davis Secretary of War* (1855), shows "THE DESERT" and "GOSHOOT INDIANS."
Courtesy Library of Congress, Geography and Map Division

westward to the Mud Lakes, and the fourth covers the extreme western edge of the Great Basin. These folded sectional maps reveal how closely the military worked with the railroad surveyors to pave the way for the iron horse. Their numerous sheets also reveal something else: The region was now becoming well enough understood that no one map could include all of the known information about it. Although the Great Basin would appear on United States maps as an identifiable region, those hoping to dominate it now began to tackle that job section by section.

As government surveyors mapped the region, they relied on a combination of modern instruments and an ages-old technique of consulting indige-nous informants. Native place names soon decorated maps, and they got there through an elaborate process of interrogation. Just after mentioning that "the mountain-range immediately to our west is called by the Indians *Pah-re-ah,* or Water Mountain, on account of the many streams which flow down its sides into Kobah Valley," Captain James H. Simpson confirmed how the process of naming occurred. "Some fifteen or twenty Diggers have come into camp," he wrote. "From these I have been enabled to get the names of some of the mountains and streams."[15] The result of this process is evident not only on the Simpson expedition's maps, but on modern maps of the Great Basin, which are festooned with names of features in Paiute, Shoshone, and other Indian languages: Beowawe, Toiyabe, Tonopah, and the Wah-Wah Mountains are all testimonials to this process of naming and claiming places on maps.

The early geologists also left their mark on the region's maps. As part of Simpson's expedition, geologist/meteorologist Henry Englemann made many significant observations. Englemann was especially fascinated by the impressive exposures of fossiliferous Paleozoic-era limestones seen in nu-merous mountain ranges in the central Great Basin.[16] By identifying and then mapping the location of these fossils, geologists began to decipher the complex geological history of the region. Consider, however, their dilem-ma. In the Great Basin, they traversed mountain range after mountain range composed of different materials. Here one might encounter sedimentary rocks like stratified limestones or sandstones in one area, metamorphic slates in the next range, intrusive igneous rocks like granites in another, and extru-sive rocks like basalts and andesites in still others. To the early geologists, the region was like a huge jigsaw puzzle, but worse; the pieces seemed to come from several *different* puzzles. From 1850 to the early 1860s, they could only patiently catalog what they observed, and draw only the most tentative of maps to suggest how the bedrock geology was configured. They often accompanied survey teams whose main goal was figuring out how to cross the infernal place safely—first by wagon, and then by railroad.

These were times of intense interest in westward expansion and resource exploitation, and the early Geological Survey—before there was in fact a federal agency by that name—was in the thick of it. Given the opportuni-ties opening up in the expanding country and the propensity of the times to encourage personal opportunism, it is no surprise that ambitious and energetic individual geologists—for example, Clarence King and Ferdinand V. Hayden—would develop intense rivalries. At this time, Congress was less concerned about battles between bright but egotistical individuals and more interested in national development, so it let those rivalries go pretty much unchecked. And why not? Working at a feverish pace, these geologists reconnoitered huge areas, deciphering nature's secrets that in turn put both precious and base metals into the treasury and the factories.

These varied survey teams caught the attention of a reading public eager to learn about the Intermountain West. Mark Twain humorously recalled a railroad survey team in western Nevada. The governor, Twain claimed, converted a motley group "into surveyors, chain-bearers and so on, and turned them loose in the desert." This, Twain observed, was "recreation on foot, lugging [survey] chains through sand and sage-brush, under a sultry sun and among cattle bones, cayotes [sic] and tarantulas." Confused about how far they would ultimately survey, the team inquired but got no definite information. When pressed, the governor urged them to keep going: "To the Atlantic Ocean, blast you!—and then bridge it and go on!" The team sent in a report and gave up surveying. The governor finally revealed that he had been joking about that Atlantic destination. Instead, he observed that "he meant to survey them into Utah and then telegraph Brigham [Young] to hang them for trespass!"[17] Twain's anecdote made light of railroad survey teams, but they got the job done.

The search for railroad routes through the Great Basin reached a fever pitch in the early 1850s. One of them, the Atlantic and Pacific Railroad Survey of 1853, charted the area in the vicinity of present-day Esmeralda County, Nevada, in search of a pass through the formidable Sierra Nevada. Geographer Paul Starrs notes that the A&P's survey was a "distinctive, if generally forgotten, chapter in Western historical geography" that resulted in several excellent maps by George Henry Goddard. Although the survey party's route never materialized as a transcontinental railroad right-of-way, Goddard's legacy was assured. He was one of the numerous railroad surveyors responsible for unraveling many of the region's cartographic mysteries as they explored routes for the iron horse. Their concerns covered not only the lay of the land only as it might offer reasonable grades, but also the availability of water and fuel for locomotives. Goddard was also an accomplished

FIGURE 6.6.

George Henry Goddard, *Map of the State of California* (1857).
Courtesy David Rumsey Collection


artist, as evidenced by the sketch that he drew of his mining claim in California. As cartographer, artist, and amateur naturalist, he met with extreme
hardship as he mapped the most punishing parts of the Great Basin. In the
fall of 1853, he collected more than six hundred botanical and geological
specimens and "mapped much new country—including, for the first time,
the upper reaches of Death Valley." His 1857 *Map of the State of California* (fig.
6.6), which included the portion of Utah Territory that would become
Nevada, "even today is widely regarded as an aesthetic masterpiece of cartography."[18] It influenced mapmakers for a generation.

At about the same time Goddard was making cartographic history, surveyors for the General Land Office also mapped huge areas of the Great Basin. Their goal was to assess land that might sustain settlers through agriculture or livestock grazing. Some of the country they traversed and surveyed was bleak even by Great Basin standards, but none more so than Death Valley. And yet even this part of the region was surveyed as possible land for settlers. In 1856, William Denton and Joel H. Brooks made a detailed map of Death Valley. In 1857, self-proclaimed "Colonel" Henry Washington extended the survey farther into the valley. These surveys helped bring some order to the chaotic wilderness of Death Valley and the Amargosa River basin. Washington's survey of the southern edge of the valley revealed "curious twin basins" that became both a cartographic icon and the source of one of the region's many colorful place names. These basins, which on the map "appear to form the eyes and face of an owl, gave rise to their current name, the Owlshead Mountains." Published in 1857 as part of the surveyor general's map of California, maps by Henry Washington and Allexey von Schmidt were better than any that would be made for the next thirty years but were generally ignored because of the surveyors' tendency to exaggerate the suitability of land for farming. As historian Richard Lingenfelter has noted, "[I]t took over three decades of further surveying to rediscover what was already known."[19]

This concept that the Great Basin was becoming known—that is, losing its mystery—had to be regarded with ambivalence in the Victorian mind. Ever attracted to both the exotic and the unexplored, the famed British explorer Sir Richard F. Burton made his way to the Great Basin in 1860-61. Two things drew Burton to the region. First, perhaps, was the Mormons with their fascinating religion that contained a hint of patriarchal Middle Eastern customs like polygamy. Second was the desert environment; this, too, suggested something of the Middle East. Combined, these two factors were a strong ingredient of a rapidly developing Orientalism—a fascination with the Near (and Far) East. This exotic land, the Great Basin, was conveniently located in the American West. In Burton's best-selling book *The City of the Saints and across the Rocky Mountains to California,* he sought to demystify both the Mormons and their new homeland on the one hand; yet on the other, as an explorer, he had a vested interest in making the region appear wild and dangerous. Having traveled through "hundreds of miles through wild country"[20] (as he called it), Burton made sure to remind readers that "Utah Territory, so called from its Indian owners, the Yuta—'those that dwell in mountains'—is still to a certain extent *terra incognita,* not having yet been thoroughly explored, much less surveyed or settled."[21]

When Burton wrote these words nearly a century and a half ago, he was aware that he was making history by describing the last North American region south of the Arctic to be fully explored—the Great Interior Basin of the American West. Like most nineteenth-century writers, however, Burton was torn by two opposing motives. He had the desire to reach an unexplored area that possessed a kind of magic precisely because it was unknown. At the same time, he also desired to inform people about the unexplored land, and hence destroy its mysterious quality by making it known. It is this conflict that makes the ongoing search for *terra incognita* such an enigma of the Age of Exploration. Just as losing one's virginity involves the benefits of "coming of age" and the liabilities of "losing one's innocence," that remarkable transition from *terra incognita* to *terra cognita* paradoxically involves both gain and loss.

Burton's descriptive prose may have given readers the impression that this land was completely unknown, but in fact it had been depicted on maps for about a century. As he traversed it by stage and wagon, Burton had in hand the latest official maps. Far from breaking new ground as he had done in Africa and the Middle East, he was in fact a latecomer here. He portrayed the region as unknown, when in actuality its status as *terra incognita* was about to end. By the time Burton traveled here, the railroad promised a quick and efficient method of crossing the region, although railroad construction was prohibitively expensive and would require additional incentives before becoming a reality. At this time, most people traveled on horseback or by horse-drawn or ox-drawn wagon. These included notables like Burton and aspiring writers like Mark Twain, as well as thousands who made the trek west to California and Oregon. Regardless of their social status, most people traveled through the region at a slow pace.

The movement of people and goods across this wilderness presented a challenge that the federal government was engaged in addressing. In 1859, Captain James H. Simpson traversed much of the region with the goal of developing better transport routes. His exploration and survey of the Great Basin from Camp Floyd to Genoa (Nevada) was conducted under the aegis of the Topographical Engineers. Numbering sixty-four persons, including surveyors, geologists, botanists, astronomers, and photographers, Simpson's expedition had "three sextants, three artificial horizons, one astronomical transit, four chronometers, two barometers, and several prismatic and pocket compasses."[22] Drawn by J. P. Mechlin to a scale of 1:1,000,000, the expedition's resulting map was supplemented by other visual information—"profiles, diagrams, and sketches"—that further revealed the geography of the Great Basin. In summarizing this expedition's accomplishments, George Montague Wheeler later observed that its report "contains original topographical data of parts of the Great Interior Basin, then (1859) but little known."[23]

Simpson's cartographic accomplishments deserve a closer look. His *Map of Wagon Routes in Utah Territory* (fig. 6.7), published in 1860, is among the most intriguing maps of the region ever produced. In addition to portraying the region's rather primitive road network, it depicts the topography in considerable detail. Its tight, crisp lines reveal the power of steel-plate technology to produce thousands of copies with little noticeable decline in quality. But the map's high quality goes beyond technology and lies in the masterful rendering of features like the Great Salt Lake, deserts, and mountains. The contrast of intense detail and white open space also helps distinguish this map as a work of art and a tribute to science.

That white space is worth exploring in more detail as it is so evocative. One is tempted to speculate that the space is left white for a simple reason—that the land was literally unknown. That interpretation, however, fails to acknowledge something more profound about such blanks or "silences": namely, that these cartographically open spaces are pregnant with meaning. In interpreting their presence on maps of Australia, cartographic historian Simon Ryan notes that such white spaces "actively erase (and legitimize erasure of) existing social, and geo-cultural formations in preparation of the projection and subsequent emplacement of a new order."[24] Tellingly, the Great Basin was aggressively mapped at the same time as Australia's huge desert region called the Outback. Both resonated as empty space to cultures bent on possessing continental interiors and displacing native inhabitants.

Then, too, the articulation of known topography further endorses the power of maps to both identify and claim space. A closer look at any mountain range on Simpson's map reveals the hachure technique at close to its

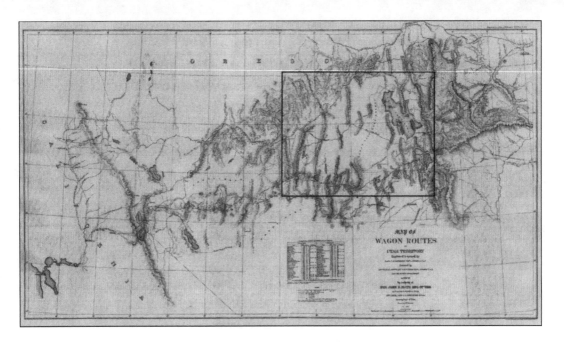

FIGURE 6.7.

J. H. Simpson, *Map of Wagon Routes in Utah Territory* (1860).
Courtesy Special Collections Division, University of Texas at Arlington Libraries

zenith—a technique that inscribes space much as a signature confirms identity. We will recall that Humboldt sensed the power of this technique as
early as 1811. Derived from the French word *hacher* (to chop up, or hash),
hachures are inscribed in the direction of the slope, drawing the eye up
and down simultaneously to create the *impression* of topographic relief. On
Simpson's map, the engraver cleverly uses darker, heavier lines to depict
those places where the relief is most pronounced. The word *hachure* enters
the popular vocabulary in 1859, at just the time this map was being prepared. The increasingly visually literate public found such maps both easy to
read and intriguing. Holding the map in one's hand, the eye marches across
the Great Basin, its journey interrupted by its beautifully articulated water-
courses and detailed mountain ranges.

Consider, too, how names help in the process of claiming places.
Although earlier maps featured *trails,* this map's depiction of wagon *routes*
suggests the region's growing accessibility. These words and symbols work
hand in hand. Whereas the earlier explorer's routes are shown by single
dashed lines that suggest their tentativeness, the wagon routes are now
depicted as twin parallel solid lines—a subliminal reference to the evenly

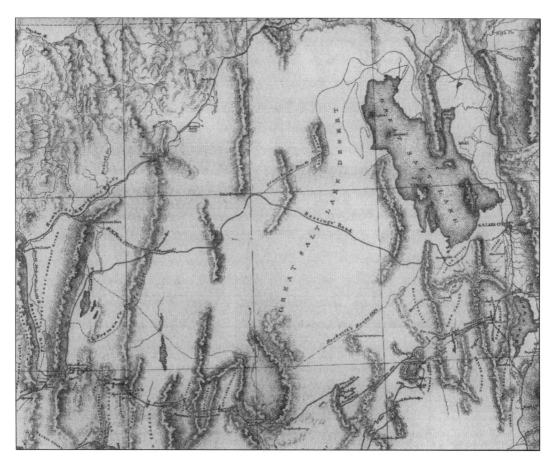

DETAIL OF FIGURE 6.7

spaced ruts left by wheeled vehicles. At just this time (1860), the fabled Pony Express route was momentarily etched into the arid landscape of the Great Basin and the process was clear: through lines on the map, the region was being drawn, quite literally, into the public domain.

Simpson's impressive *Map of the Shortest Route to California* (1859) not only helped to lay the groundwork for improved roads; it also reveals a growing awareness of the region's internal geography as it depicts the topography and hydrology in increasing detail. Although the compilation of Simpson's map depended in part on works of earlier cartographers, including the fabled Frémont-Preuss maps of 1845-48, his 1859 effort breaks ground. Maps of this type helped guide writers like Mark Twain and Richard Burton across the region to California, but they also helped encourage miners who began flocking to the area in search of precious metals, in part a result of the California gold rush. During the 1850s, prospectors explored many portions of the region. Some of these prospectors had been over the same ground once before when they first crossed it on the way to California. Returning with new knowledge, they discovered the fabled Comstock Lode at Virginia City in the late 1850s. By 1859-1860, the mining-inspired Pyramid Lake War had severely affected the area's Native American inhabitants in present-day western Nevada. In Utah, at the eastern edge of the Great Basin, the relationship between Mormons and Indians was more complex. The Latter-day Saints found themselves fighting some Indians as Mormon settlements expanded, and missionizing others in the belief that all Native Americans were Lamanites, or members of the lost tribes of Israel.

By the late 1850s, the region's internal geography was becoming much better understood by an increasing number of expeditions that in turn mapped more of it and made these maps available to a public eager for information. Then, too, the Mormon presence, and the Saints' search for additional places to settle away from growing governmental persecution, led to increasing knowledge about the region's geography. In settling Mormon Deseret, however, Brigham Young relied on any accurate maps he could find, including those like the Frémont-Preuss map, which was produced by a federal government he increasingly mistrusted. The Mormons also made many of their own maps to help them explore and settle portions of the Great Basin. This subject, the role of Mormons in mapping the West, reveals that the West they claimed had a presence on paper before a street was graded or a ditch dug. In their aggressive proselytizing, the Mormons sought talented people from as far away as Europe, and surveyors/mapmakers were among them.

Young also sought cartographers closer to home. From within the ranks of church members, talented mapmakers emerged. Some, like Thomas Bullock, not only drew maps of routes that could take the Saints into the far West but also drafted some of the earliest detailed maps of portions of the Great Basin. In fact, Bullock's August 1847 plat map of the new "City of the Great Salt Lake" is the first known map of a settlement in the region. Bullock also assisted church authorities by mapping routes from Utah into the Indian villages of Arizona in the late 1850s and early 1860s as the Saints expanded their missionary efforts into Hopi country. Other Mormon cartographers such as George Washington Bean drafted numerous maps intended to identify the most effective routes through the Great Basin as the Latter-day Saints endeavored to settle the entire region as part of the huge theocratic Kingdom of Deseret. Still others, like the modest but talented John Steele, also helped the church map the trails, topographic features, landmarks, and settlements in areas like the fledgling settlement at Las Vegas.[25] Considering the Mormons' ambitious recruitment and colonization efforts, it seems surprising that virtually all of their maps remained unpublished—until we remember the political situation of the mid-to-late nineteenth century: Given the suspicion with which Mormons viewed critical outsiders and increasingly hostile federal authorities, it is no surprise that they closely guarded their maps. The information on these maps was a two-edged sword. It could help the Saints expand their hold on the Interior West, but if it fell into the wrong hands, it could just as easily be used to help outsiders displace them. Drawn at the same time that the Mormons even created their own alphabet, these maps helped members of the religion to interact more or less secretly.

Consider the case of maps that could help Mormon leaders pursue the most secret of strategies that would enable the church to survive on the eve of the Utah War. As tensions between Mormons and "Gentiles" mounted and the United States planned to take Salt Lake City by force, the Mormons prepared to flee farther into the desert—which is to say deeper into the Great Basin—for sanctuary. The rugged desert-mountain region presented great challenges, and they needed maps to chart the way. James H. Martineau's *Chart Showing the Explorations of the Desert Mission* (fig. 6.8) and Orson B. Adams's *Map of the Desert, June 1858* confirm that the Mormons were actively mapping large parts of the region as conditions demanded.[26]

These maps reveal a developing articulation of the area following the Frémont-Preuss mapping a decade earlier. Clifford Stott observes that the Pathfinder's reports and maps "were given wide credibility" by the Mormons.[27] However, it should also be restated that the geographic information provided by the Mormons found its way back to Salt Lake City and into church headquarters. The result of this Mormon reconnaissance was that maps of the region developed more or less along two lines—those used by the Latter-day Saints, and those used by civil authorities. In late-1850s Utah Territory, these two forces were often distinct and opposed. Their maps revealed a rapidly growing knowledge of the region that would find both

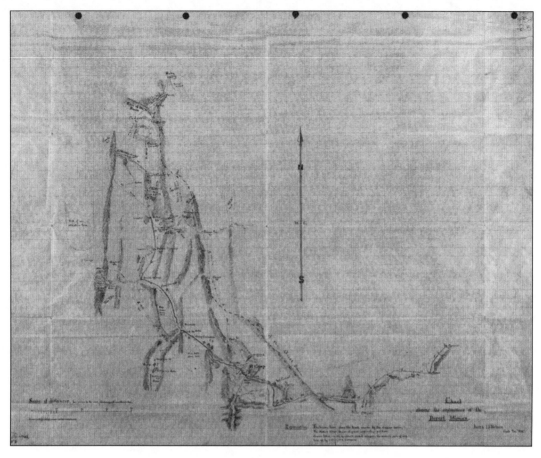

FIGURE 6.8.

James H. Martineau, *Chart Showing the Explorations of the Desert Mission* (1857).
Courtesy Archives, Church of Jesus Christ of Latter-day Saints

justifying claims—one religious, the other secular. And yet, by the 1860s and 1870s, when the federal government surveyed portions of Utah, they employed talented Mormon cartographers like James Martineau—another reminder that mapmaking sometimes involves unexpected partnerships.

Mapmaking is a solemn activity, but there were moments of levity as maps of the region took shape. The rather serious issue of place names sometimes took a humorous turn with so many individuals and groups at work. In 1863, after he drafted *Bancroft's Map of the Pacific States,* cartographer William Henry Knight was chided by a friend for omitting the name of Lake Bigler. Knight had left the lake unidentified because its namesake—a former governor of California—had fallen out of favor. However, having anticipated such criticism, Knight responded that he had a list of several

Indian names that could replace Bigler's. Knight liked Tahoe the best, and so Lake Tahoe it became.[28] Based on as much information as his sponsor Hubert Howe Bancroft could provide, Knight's map was both beautiful and informative, a stunning combination of information from public and private sources. That considerable debate raged surrounding the naming of this lake is evident from the newspapers of the time. In 1864, the *Reese River Reveille,* published in the booming mining town of Austin, Nevada, noted disagreement as to what the lake should be called. While some preferred Lake Bigler and others Lake Tahoe, a third option now appeared. "On maps in France showing this part of the country," the editor noted, "the lake is given the name Bonpland." This possibility certainly possessed some merit, for Amande Bonpland, a member of Frémont's expedition, was said to have discovered the lake in 1844, and Frémont himself urged that it be named after the French botanist. Actually, Bonpland had something else in his favor: He was the companion and friend of Baron von Humboldt, whose name— thanks largely to Frémont—still graces several geographic features (the Humboldt River, Humboldt Sink, and the Humboldt Mountains) in Nevada. The *Reese River Reveille* had a simple solution. "Let the name Bonpland remain on the maps in France," the editor opined. "We can get along with that of Tahoe," he concluded, because that name was "of Indian origin and more antiquity than Bonpland, and every way more suitable."[29]

# 7
## Filling in the Blanks
### 1865–1900

BY THE MID-1860S, POLITICAL REALITIES and divisions further shaped both popular perceptions and the mapping of the Great Basin. Before that time, the region had been part of one larger political entity—for example, Upper California (until the 1840s), Mormon Deseret (late 1840s to the early 1850s), or broadly, Utah Territory (in the 1850s and early 1860s). Now, with the creation of Nevada (1864), the region exhibited a longitudinal (that is, east-west) split into two separate political entities. Drawn on one of the meridians, that prominent north-south state line effectively bisected the region into one state and one territory in the late nineteenth century.

In the popular mind, Nevada came to be associated with mining, and Utah with agriculture. The mining of precious metals—an activity condemned by Brigham Young in numerous sermons—had an especially powerful effect on the region and maps of it. The prospect of rich mines helped motivate the U.S. government to conduct detailed geological study. In 1867–72, Clarence King's Geological Exploration of the Fortieth Parallel had been authorized "to examine and describe the geological structure, geographical condition, and natural resources along that parallel between 105° and 120° longitude, with sufficient expanses north and south to include the lines of the Central and Union Pacific Railroads, and as much more as may be consistent with accuracy and a proper progress." Through successive seasons, the exploration teams were commanded to "examine all rock for-

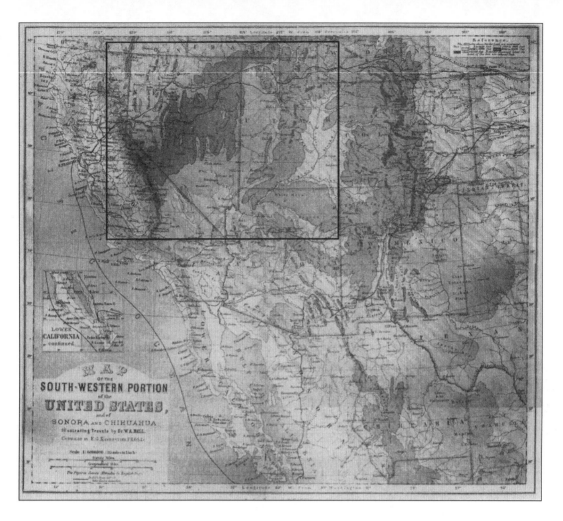

FIGURE 7.1.

E. G. Ravenstein, *Map of the Southwestern Portion of the United States and of Sonora and Chihuahua* (1867), from William A. Bell's *New Tracks in North America: A Journal of Travel and Adventure whilst Engaged in the Survey for a Southern Railroad to the Pacific Ocean in 1867–1868*. Courtesy DeGolyer Library, Southern Methodist University

mations, mountain ranges, detrital plains, mines, coal fields, salt basins, etc., as well as also for a topographic map of the region traversed."[1] The goal was as much economic as scientific. The region's topographic fieldwork was accomplished using Zenith telescopes to determine latitude, a Zenith sextant and chronometers to determine longitude, a 4-inch gradienter for topographic details, steel tapes and chains, and cistern barometers (to determine elevations).[2] These surveys helped confirm what many prospectors already knew—the region's mountain ranges were treasure troves of metallic ores waiting to be exploited.

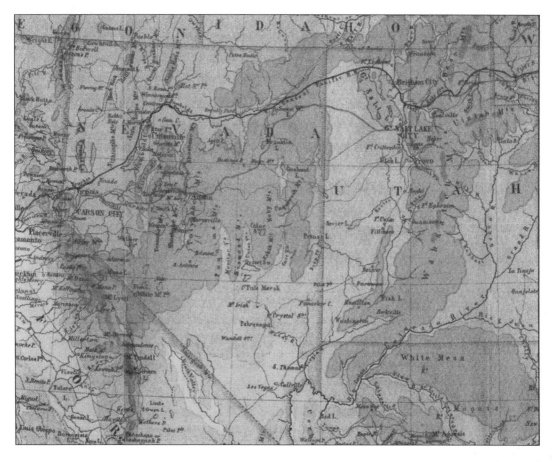

DETAIL OF FIGURE 7.1

In the last half of the nineteenth century, the region came under scrutiny by the scientist explorer. These individuals had a long tradition, dating
from the Spanish scientists as well as Alexander von Humboldt. However,
they now appeared in relatively new form associated with America's westward movement. In the Great Basin, we can trace their roots to the early
pioneering expeditions of scientists like Stansbury, who interpreted and
illustrated the area adjacent to the Great Salt Lake in the early 1850s. These
expeditions often featured geologists in key roles. The American Civil War
(1861-1865) momentarily slowed the scientists' inquiry, but their scrutiny
was irrepressible. This can be seen in the second flowering of geognostic
study following the war's end. By 1867, when E. G. Ravenstein compiled his
*Map of the Southwestern Portion of the United States and of Sonora and
Chihuahua* (fig. 7.1), the interior Great Basin was best known along the well-
traveled Humboldt River corridor, but the elevation of many points else-

where awaited determination. The hypsometer, a device for estimating the elevations of mountainous locales based on the boiling points of liquids or by triangulation, helped in mapping the region at this time. Ravenstein's map is among the earliest to indicate the lay of the land hypsometrically— that is, to depict the topography in various categories by elevation, each category colored/shaded to stand out boldly. Ravenstein's map shows a broad elevated area in central Nevada but does not depict the numerous mountain ranges in the southern Great Basin. Instead, it suggests that this area is a plain sloping toward Las Vegas. Here and there, that sloping surface is dotted by springs and punctuated by a mountain peak.

Significantly, Ravenstein does not depict the mysterious transverse mountain range that maps by Frémont-Preuss and others had shown in the 1840s and 1850s. In the book to which Ravenstein's map is appended, William A. Bell seems to lay to rest forever that mythical feature: "There is nothing whatever in the physical construction of the Great Basin to have prevented the formation of one great river, emptying into the Gulf of California, either as an independent stream or as a tributary of the Rio Colorado."[3] Intent on shattering myths, Bell went on: "It is not because the Great Basin is really a complete basin without an outlet, or with a rim presenting an insurmountable barrier to the drainage" that its waters do not reach the sea. Rather, Bell explained, "it is not a single basin at all, but a collection of perhaps hundreds of basins, which have remained in their primitive isolated condition . . . because the separate streams never had enough force to break through the barriers . . . to form a complete draining system."[4] Even though Bell realized that "more rain fell" here in the past, he did not know that the region had drained northward into the Snake River during Pleistocene times. Bell was writing at a time when geologists were rapidly learning the secrets of the Great Basin's past, and those scientific discoveries were often a few years ahead of the popular press.

In reflecting back on 1867, Clarence King astutely recognized the year as a watershed: "Eighteen sixty seven marks, in the history of national geological work, a turning point, when the science [of geology] ceased to be dragged in the dust of rapid exploration and took a commanding position in the professional work of the country."[5] This professional work, as he called it, was of immense interest to the public. No one was in a better position to fathom this than King, one of geology's remarkable figures—equally at home in the field mapping and speculating about the geomorphology, in the laboratory analyzing specimens, or standing before Congress requesting funding for the U.S. Geological Survey. The geological and cartographic knowledge of the Great Basin made impressive strides under King, whose *Report of the Geological Exploration of the Fortieth Parallel* reveals an ability to map not only spatially, but also dimensionally. Of the five volumes, the first three deal with, respectively, the region's systematic geology, descriptive geology, and mining industry. These volumes reveal both the region's geological complexity and the concerted efforts to decipher it using new techniques. King provided two types of maps. In the individual volumes, he used black-and-white images, while the accompanying atlas-plates made stunning use of what an art historian calls "the democratic art"—chromolithography.[6] Both types of images show how geologists conceptualized the region and used cartography to depict the varied geology through horizontal sections and cross sections. The volume entitled *Mining Industry* presents cross sections of particular mining districts, such as the Washoe (fig. 7.2). As indicated on horizontal sections, geologists drew lines from various points, much as a surgeon would mark the line of an incision. Then, in cross sections, they revealed how the interior of each section would look if viewed from the side. By viewing such maps, we become privileged insiders, literally and figuratively.

Much like a carefully excised slice of plant or animal tissue, these sections were essentially anatomical and, as such, become part of a narrative about both functions and origins. King was fairly explicit about the power of the "Washoe section," which helped him illustrate the significance of extrusive vulcanism. This section, he noted, permits the reader of both text and map to understand "evidence of a long, complicated geological history." As King put it, "It is rare to find within such a small area all the representatives of the volcanic family, and with them the relics of one of the ancient ranges." But that is exactly what the reader of his report and his map is privileged to experience. King continues: "The reader, by comparing these [sections] with the position of the rocks on Atlas-Plate II, will easily perceive the general geological relations of the district."[7] Later in this same volume, King discusses and illustrates the geology and mines of central and eastern Nevada. To provide a better understanding of the complex geology with its numerous faults and discontinuous ore bodies, he presents cross sections of

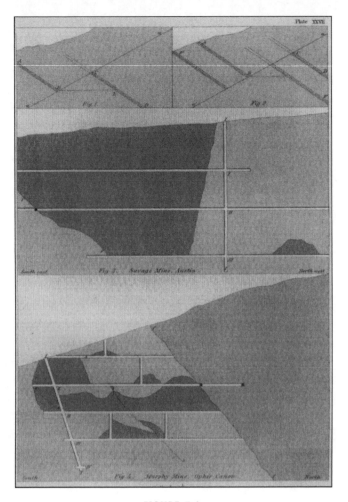

FIGURE 7.2.

Cross sections of the geology in the Virginia City, Nevada, area from
Clarence King's *Mining Industry* report (1872).
Courtesy Special Collections Division, University of Texas at Arlington Libraries

several mines in the vicinity of Austin, Nevada (fig. 7.3). Note that the varied mineshafts and cross shafts are indicated—a suggestion that mining and increasing knowledge go hand in hand. The faults show the stark contrast between promising/productive areas and barren areas of nonmineralized "country rock." These cross sections are more than simple statements about economic motives; they reveal the increasing intimacy with which the earth was explored. As always, increasing intimacy coincides with demystification—the cost, as it were, of learning previously hidden secrets. Clarence King intuitively understood that such acts of description and mapping meant a sea change in perception and authority.

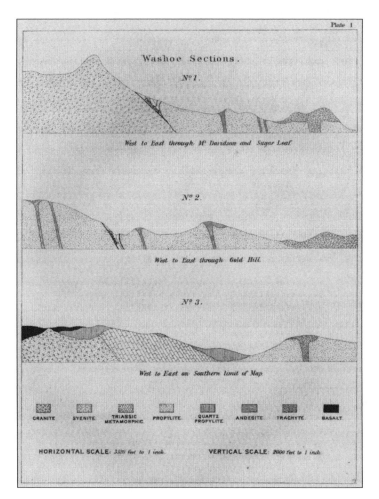

Plate 1

Washoe Sections.

No 1.

West to East through M? Davidson and Sugar Loaf

No 2.

West to East through Gold Hill.

No 3.

West to East on Southern limit of Map

GRANITE   SYENITE.   TRIASSIC
METAMORPHIC.   PROPYLITE.   QUARTZ
PROPYLITE   ANDESITE.   TRACHYTE.   BASALT.

HORIZONTAL SCALE: 3520 feet to 1 inch.   VERTICAL SCALE: 2000 feet to 1 inch.

FIGURE 7.3.

Cross sections of mines in the Austin, Nevada, area from Clarence King's
*Mining Industry* report (1872). Courtesy Special Collections Division,
University of Texas at Arlington Libraries

In 1876, King applied the name "Nevada" to a Paleozoic formation
shown on his atlas map.[8] The geological maps in King's atlas were among
the earliest thematic maps (that is, maps showing distributions of specific
features or phenomena) in the Great Basin. To the public, geologists like
King were both heroic and frightening figures. They not only revolutionized cartographic history in that they drew cross sections of the topography;
they also helped demystify the secrets of the earth. Their cross sections were
in effect three-dimensional diagrams that showed how the various strata
were configured underground where only the imagination or the arcane
argot of miners once reigned. The geologists on these expeditions soon real-

ized that the region's mountains were storehouses of geological history. A geologist later observed that "the Great Basin was one of the world's great repositories of late Precambrian and Paleozoic sediments."[9] Another confirms that early geologists like Englemann and King had found the perfect landscape in which to decipher and map the geology because "nowhere in western North America are fossil-bearing middle Paleozoic rocks better shown than at Antelope Valley and in adjacent territory" in eastern Nevada.[10] These textbook exposures had intrigued Englemann and King, who knew that these rocks had been deposited in seas several hundred million years ago and were now stacked in shattered, mountain-sized blocks. However, they could not yet fully comprehend the tectonic forces that had shaped the spectacular topography here. That would await revelations about continental drift a century later.

And yet, by the late nineteenth century, geologists had solved one mystery—the Great Basin's hydrologic past. By studying the striking ancient lake terraces on hillsides and mountainsides, they deduced that two huge lakes once existed here—one on the eastern side of the region, the other on the western side. The Great Salt Lake and Utah Lake are remnants of the former, and Pyramid Lake and Walker Lake remnants of the latter. In honor of those early cartographers who speculated about hydrological features here, they named these long-vanished ice-age lakes Bonneville and Lahontan, respectively. Their appearance on scientific maps in government reports helped bring the geological past alive and underscored the region's present-day aridity.

It is only in the context of such aggressive scientific mapping that we can understand the burning ambitions of a man who left both his name and his mark on the cartographic history of the region—George Montague Wheeler. To Wheeler goes the credit of deciphering the southeastern portion of Nevada, which in turn helped clarify the geography of the entire southern Great Basin. Wheeler's *Map Showing Detailed Topography of the Country Traversed by the Reconnaissance Expedition through Southern & Southeastern Nevada* (1869) is one of the region's cartographic masterpieces (fig. 7.4). P. W. Hamel served as chief topographer and draughtsman on this elaborate map drawn at one inch to twelve miles. From the small farming oasis of Las Vegas, the map sweeps northward in a breathtaking display of mountains alternating with desert valleys. As we might expect of a map by Wheeler, who championed the development of mines here, the map is festooned with postage-stamp-like rectangles depicting numerous fledgling mining districts. The map's artistry is enhanced by its use of a combination of hachures and minute dots, a technique called stippling, to depict the topography.

Like other mapmakers in this region, Wheeler knew that map users needed names as well as images to make sense out of wild places; accordingly, he took care in naming the places that he mapped, placing those names strategically as one fills in the blanks of a quiz. In addition to Indian names such as the Timpahute Range and the Pahranagat [Mining] District, Wheeler's map also identifies a Coal Valley and Railroad Valley. Somewhat ambitiously, the map also shows a "Proposed Railroad" line running from just east of Callville on the Colorado River to the Central Pacific line at Toano. This railroad line was never built, but its legacy—the name Railroad Valley—still remains today and confirms the aspirations of entrepreneurs who hoped to open the southern Great Basin to commercial development. Like other maps, Wheeler's represents not only actual fieldwork, but also the efforts of earlier cartographers who left their mark. Three, including surveys of the Central Pacific railroad, the USGS, and Lieutenant Joseph Christmas Ives along the Colorado River, were consulted. Thus Wheeler's reconnaissance was incremental, adding the crucial pieces of a puzzle that help complete the whole. At this date, 1869, only the southwestern portion of the region remained something of a mystery.[11]

Of the many government surveyors who traversed the Interior West, few are more enigmatic than Wheeler. Wheeler's biographer Doris Ostrander Dawdy observed that his work "was not a survey, it was a reconnaissance," and that much of his effort "had been primarily on mining districts, their access to transportation routes, and the disposition of the Nevada Indians."[12] Significantly, Wheeler was caught up in the internal political intrigue and backbiting that characterized much western exploration. He conducted additional reconnaissance as part of the 1870 survey under General E. O. C. Ord but secretly hoped to gain fame and upstage his competitor Andrew Atkinson Humphreys. Dawdy notes that Wheeler was surreptitiously engaged in developing his own mining interests while simultaneously conducting his fieldwork for the U.S. government—what we would today label a conflict of interest. Wheeler had an irascible personality and criticized the very people—geologists—who could provide information about the areas

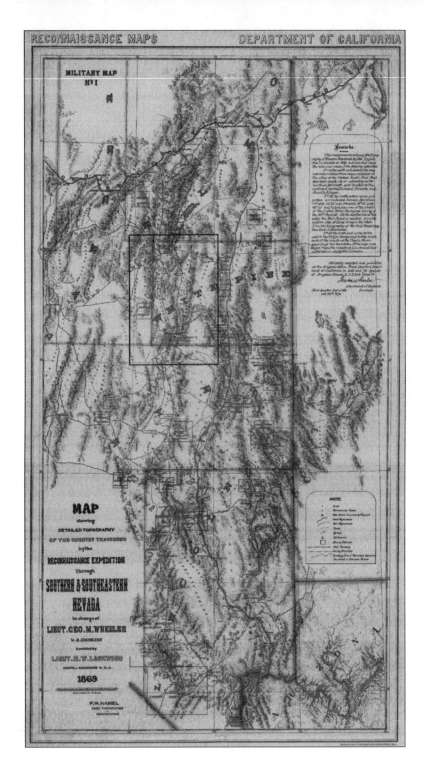

FIGURE 7.4.

George Wheeler, *Map Showing Detailed Topography of the Country
Traversed by the Reconnaissance Expedition through Southern & Southeastern Nevada* (1869).
Courtesy David Rumsey Collection

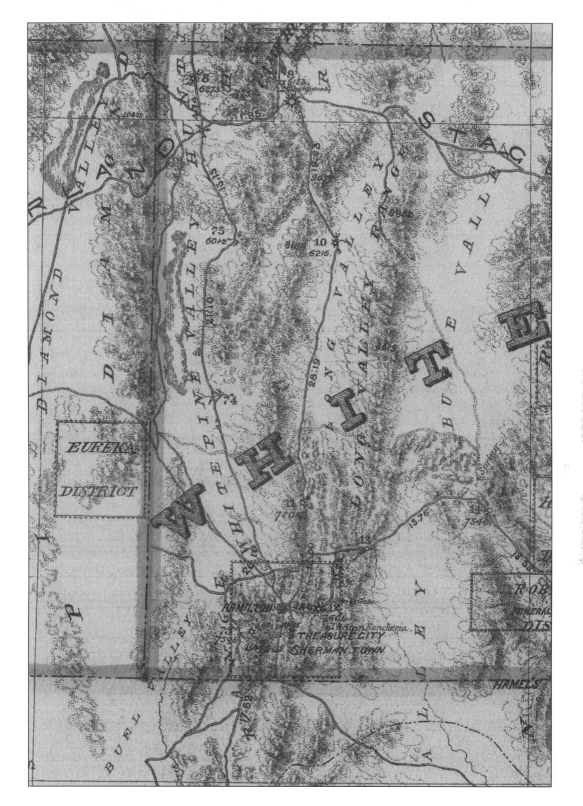

DETAIL OF FIGURE 7.4

traversed. He haughtily wrote that "the natural and necessary inferiority of all topography directed by geologists . . . arises from the condition that the inherent requirements of geological investigation demand more attention to natural features of a given area." Adding fuel to the fire, Wheeler feistily declared that geologists were "not capable or skilled in the exact or mathematical sciences that must be employed in the field observations and map construction" necessary to delineate an area.[13] Wheeler's claim is ironic, because historians are divided on the scientific value of his own surveying and mapping, and on the real reasons why Congress terminated his survey. In *Great Surveys of the American West,* Richard Bartlett suggested that Wheeler's surveys had improved and warranted continuation instead of termination. William Goetzmann, on the other hand, believed that Wheeler's failure to use the latest scientific mapping techniques justified Congress's termination of his survey. For his part, Wheeler was slow to adopt new mapping techniques—notably contour mapping, which the geologist Clarence King had mastered and used so widely. Wheeler had made one such map "but, perhaps for budgetary reasons, produced no more."[14]

Wheeler's tendency toward self-promotion irritated many, and one example of it permanently affected one of the region's place names. When Wheeler's 1875 edited version of the 1869 Nevada reconnaissance was published, it brazenly identified Nevada's highest point as Wheeler Peak. This expropriation concerned Col. James H. Simpson, who as a captain in the Topographical Engineers had discovered the peak in 1859, originally naming it for Jefferson Davis and then Union Peak when Davis joined the Confederacy. As if to infuriate everyone involved, Wheeler asserted that "[t]his peak has been called indiscriminately on published maps Union or Davis Peak." Possibly taking advantage of this confusion, he simply renamed the peak after himself. When Simpson confronted him, Wheeler agreed to change it back to Union Peak. However, as time has shown, he reneged on that promise. The tallest peak in the Great Basin remains Wheeler Peak to this day.[15]

Standing at a lower elevation in the shadow of Wheeler Peak today, Jeff Davis Peak is a mute reminder of the battle of words, or rather names. Wheeler's vanity should not detract from his passion for deciphering the complex topography of the southern Great Basin. In the summer of 1871, the indefatigable Wheeler explored much of southern Nevada but met his match in Death Valley, where he lost two guides, who disappeared in the fur-nace-hot desert while on errands and were never found. For this and other problems that plagued the expedition, Wheeler was condemned by the fledgling *Inyo Independent* newspaper. Coming to Wheeler's defense, Captain Harry Egbert, commander of Camp Independence, astutely noted that "it is exceedingly difficult to obtain appropriations from Congress to survey and examine countries supposed to be desert." Egbert's defense is also some-thing of an understatement. The country Wheeler found himself in was the epitome of aridity, its stark mountains looming out of shimmering valleys veneered with shifting sand dunes and burning salt flats. Despite these dif-ficulties, Wheeler's survey of this particularly inhospitable section of the Great Basin resulted in additional accurate mapping.[16]

It is here that we confront an important issue affecting the cartographic history of the Great Basin—namely, the difference between mapping and surveying. Whereas mapping delineates the countryside as it is thought to exist or may very well exist based on actual observation, surveying goes fur-ther. As the act of literally drawing lines to connote specific territorial divi-sions, surveying is concerned with boundaries. These boundaries may sepa-rate one political jurisdiction from another or one landowner's property from the property of another, but the effect is the same. Survey puts a stamp of approval and control on both the landscape and the imagination. We in-tuitively understand this process as one of increasing validation of claims that may have been made, albeit vaguely, on earlier maps.[17]

Among the most publicized expeditions to the region, George Wheeler's 1873 traverse not only conveyed the government's authority and the private sector's claims; it also left a series of stunning maps that substantiated specif-ic claims to specific places. Equally important, Wheeler's expedition through the midsection of the Great Basin helped fill in a number of the cartograph-ic blanks that existed up to that time. Wheeler's survey team delineated the topography, confirming the basin and range quality of the countryside. Significantly, this expedition further dispelled the long-held belief, traceable to the 1840s, that a huge range of mountains stretched from east to west at about 39° north latitude. Wheeler's expedition coincided with the national financial panic of 1873, which triggered widespread economic depression and falling metals prices. Still, prospectors combed the area, further helping to demystify it—and hoping to become rich in the process.

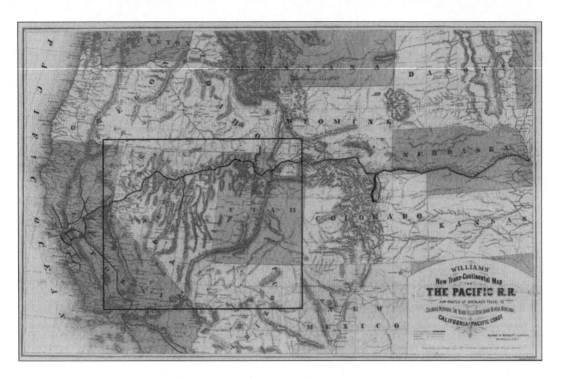

FIGURE 7.5.

*William's New Trans-Continental Map of the Pacific R.R.*
*and Routes of Overland Travel to Colorado, Nebraska, the Black Hills, Utah,*
*Idaho, Nevada, Montana, California, and the Pacific Coast* (1877).
Courtesy Special Collections Division, University of Texas at Arlington Libraries

By 1869, when the nation's first transcontinental railroad cut through the region, the central Great Basin was well known cartographically. It became especially familiar in the popular press as the stage upon which frontier drama was still unfolding. By rail, it attracted observers interested in scrutinizing the Mormons building their Zion, ranchers developing spreads on private land or using public lands for grazing, Indians trying to make do in a changing world, and prospectors bent on locating precious metals in its numerous mountain ranges. Significantly, almost exactly where Miera's 1776 map shows *terra incognita,* maps of the 1870s still showed one large, desolate area west of the Great Salt Lake, known colorfully as the "Great American Desert." This wording is an appropriate epitaph to the region as a desolate wasteland. It resonated as dangerous, and increasingly romantic, country to those who viewed it from the windows of a passing express train.

By the 1870s, the Central Pacific's portion of the transcontinental railroad line, which traversed much of the Humboldt River's course, was the route

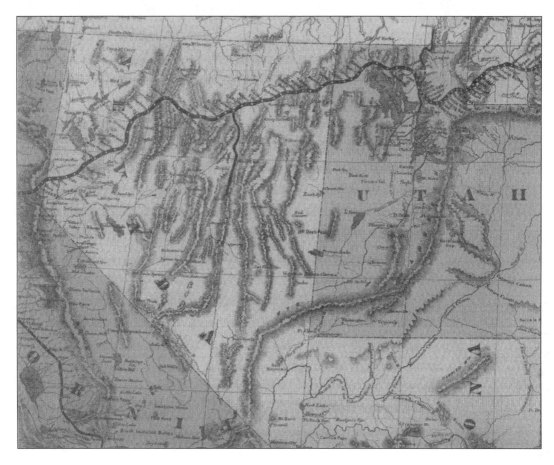

DETAIL OF FIGURE 7.5

of choice for travelers. The railroads supplemented their original surveys with elaborate maps intended to entice travelers. Products of the Victorian era, with lavish illustrations and flowery texts, railroad maps became one of the characteristic icons of the late nineteenth century. Typical of the ornate railroad maps of the period is *Williams' New Trans-Continental Map of the Pacific R.R. and Routes of Overland Travel to Colorado, Nebraska, the Black Hills, Utah, Idaho, Nevada, Montana, California, and the Pacific Coast* (1877) (fig. 7.5). Published in New York, it provides a dramatic overview of the route across the Great Basin. Although intended to show the railroads, Henry T. Williams's map does so by emphasizing their relationship to topography. Here Williams builds on information in government surveys, which were increasingly accurate. But Williams also takes some liberties with the topography as he stimulates the public's imagination to ponder the obstacles that the railroad had to either avoid or conquer. Consider, for example, his depiction of the Wasatch Mountains, which boldly snake southwestward to define the Great Basin's southeastern boundary (when in fact the topography here is considerably more broken), and mountain masses isolated at the region's watershed with the Colorado River.

Williams's map is a tribute to the railroad. On it, the Central Pacific line stands out boldly, as does the Eureka and Palisade narrow-gauge railroad line from Palisade to the mining center of Eureka. In addition to transcontinental traffic, the railroads built lines to mining towns like Eureka, and later Austin and Ely. Maps like these did much to cement an image of the Great Basin in the popular mind as a rugged, isolated desert region dotted by points of prosperity that were in turn strung out like beads along meandering transportation routes. Williams knew that railroads and mining went hand in hand. In 1870, British traveler/writer William Bell noted that "the sterility of these regions is not an unmitigated evil to the railroad that crosses them; for the miners, whose wants are very great, require all the necessaries and many of the luxuries of life to be carried to them by rail."[18] Like most travelers of the time, Bell took the Central Pacific. About a decade later, another British traveler/writer, Phil Robinson, also commented on the desolate landscape and the forlorn quality of towns strung along this same railroad line that threaded its way through the Great Basin.[19] Yet, as rendered on the Williams map, the railroad seems in perfect control, its bold black line connecting numerous communities bearing mysterious Native American names—Tulasco, Be-o-wa-wa, Shoshone—although in reality most of these were small tank towns and little more. The map beautifully reveals how the Central Pacific wove its way across the region following the easiest grades along the Humboldt River, missing those bold, hachure-defined mountains that look like an army of wooly worms frozen in their north-south march.

With railroads, the sustained production of the mining districts was assured. The mapping of individual mining districts forms yet another chapter in the cartographic history of the Great Basin. That chapter had opened with the development of the silver mines at Virginia City after 1859. By the time Samuel Bowles traversed the region in the late 1860s, he could observe that "Nevada's claim to the name of the Silver State is not only good yet brightening." Bowles's optimism clearly sprang from the development of mines all across the state. "How well these later mining discoveries and

developments are distributed over the broad area of the State," he concluded, "will be impressed on every student of the map."[20] The map that accom-
panies Bowles's book revealed that Virginia City had company by then, as
Austin and Parranagat were also shown, with asterisks indicating their
importance. Bowles's small-scale map does not show these as mining towns
per se, but his text describes Nevada's new mining areas in considerable
detail.

How, one wonders, did Bowles acquire this information, and why did he
allude so pointedly to "the map"? Bowles's map and text were likely inspired
not only by his travels but also by an "official" source—the *Map of the State
of Nevada* (fig. 7.6) by the U.S. Department of the Interior, General Land
Office, which accompanied the *Annual Report of the Commissioner of the
General Land Office in 1866*. Readily available to enterprising speculators, this
superbly drafted map is part of a series, and it matches the California map
prepared at the same date and scale. The year 1866 is important, for the
Great Basin was rapidly losing its status as *terra incognita*. Of special interest
to mining historians is the identification of three types of mining areas (generally called "districts")—for silver, gold, and copper. The map's color cod-
ing depicts gold mines in yellow, silver mines in blue, and copper mines in
green—although the two latter colors are so similar that they are rather difficult to differentiate. At the date of publication, the Central Pacific railroad
was still "proposed," but its route is shown pretty much as it would be completed within three years. Topography is depicted in relation to elements
that can sustain life, such as a "Mountain Range whit [*sic*] but little wood,
grass or water" near Pyramid Lake or another "Covered with Nut Pine and
Juniper" in southern Esmeralda County. The map also depicts other geographical or geological features—for example, "Hard White Clay" in
Esmeralda County and "Coal Signs" in northern Nye County. Note that
many of Nevada's famed mining districts—the Comstock, Reese
River/Austin, Aurora (the latter erroneously depicted as being in California
on some early maps!)—are shown, but the area around Tonopah is labeled
merely as "Desert without wood water or Grass [*sic*]" and as the site of a
"Soda Spr[ing]." Although the town of Silver Peak began to develop as early
as 1865, it is not shown. It would take another third of a century for this
seemingly barren area to become the site of Nevada's last big bonanza—the
Tonopah/Goldfield rush of ca. 1900-1907. This map nicely illustrates the

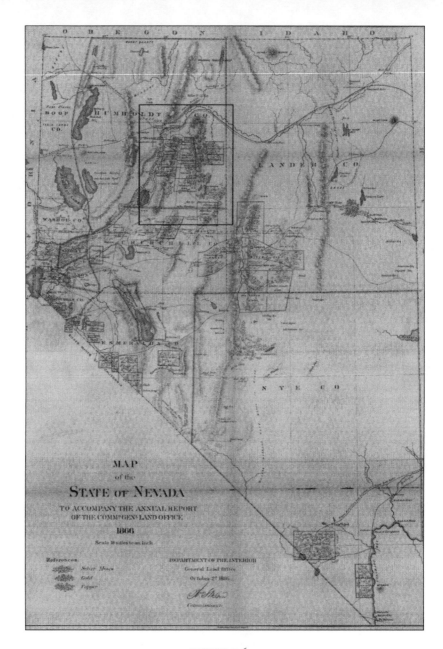

FIGURE 7.6.

General Land Office, *Map of the State of Nevada* (1866), showing mining districts.
Courtesy Special Collections Division, University of Texas at Arlington Libraries

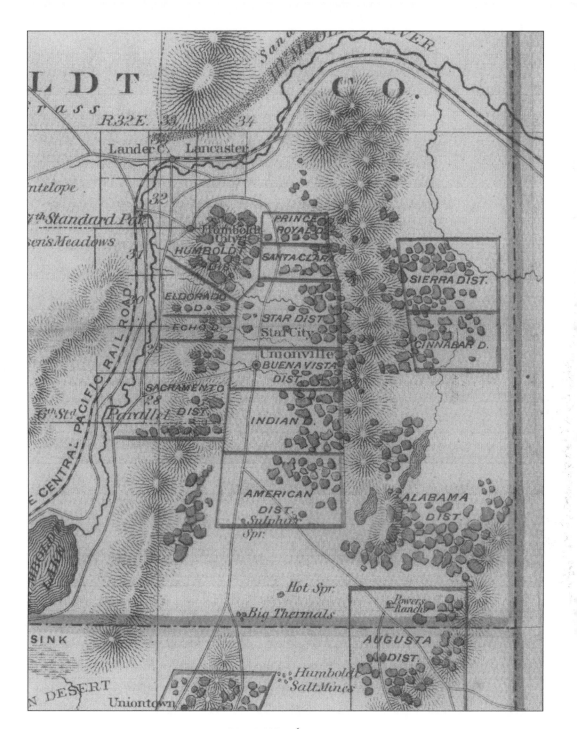

DETAIL OF FIGURE 7.6

mapping of mining at a relatively small scale (hence showing a large area) to place activities in broad geographic context, but of course such a map cannot show individual mining districts in much detail.

Consider the challenge facing those who wished to precisely locate a mine in such a vast region. First, they needed to determine the context of the mine in relation either to familiar landmarks or to previously surveyed points. Recorders of claims in mining districts had to think both legally and spatially. A historian perceptively noted that "the recorder was the key man in the mining district." A description of Nevada's Shoshone Mining District underscores the challenges of determining location. The district, part of Township 11 North, Range 68 East, is on the west slope of the Snake Range. As the Nevada state mineralogist described it in 1870:

> This district joins Lincoln on the south and has all the same natural facilities for mining. The mines . . . are situated on a low spur of the mountain called Mineral Hill. Another spur further north called Lookout Mountain has a number of mines. East of these hills is a cañon, at the head of which a saddle connects the hills with the main mountain. This saddle rises into another ridge known as the Hotchkiss Hill. North of this there is a wide cañon in which a village is surveyed. North of this cañon there is a bench or level place on the top of a hill known as Bromide Flat, where there are mines. Nearly the whole space described is covered with nut pine and mountain mahogany. To the east, the mountain rises very high, probably ten thousand feet, and capped with limestone.[21]

The key to determining location accurately amid such remote and convoluted topography was the survey, but surveying was conducted at many levels. In the case of mines in the Intermountain West, accurate local surveying was built upon the U.S. government's recently completed general survey, which placed a particular location in reference to a developing national grid of baselines and meridians. Thereafter, local surveyors' maps formed the basis for not only delineating individual mines, but also mapping entire mining districts.[22] Consider the daunting task facing cartographers mapping a small but complex mining district. Their work is exacting, for the maps they produce will be used to locate particular mines, and also to avoid disputes as claims frequently adjoin each other. Thus it is that a new type of map—one highly detailed and depicting individual claims in proximity to one another—marked a new phase in the cartographic history of the region. These maps depended entirely on good surveying—another reminder that the region was being brought into the fold of the expanding nation.

These maps were the result of aggressive scientific interest by government geologists and equally aggressive investment by mining promoters in the private sector. By the 1870s, the region was the domain not only of the scientist and entrepreneur, but also of the bureaucrat. In 1878, explorer and self-made scientist John Wesley Powell oversaw the production of *Map of Utah Territory: Representing the Extent of the Irrigable, Timber, and Pasture Lands.* Its cartouche lists the names of the four men who compiled and drew the map, another reminder that maps are the result of teamwork. Like the geologists whose maps served as inspiration, these cartographers produced a thematic map that would soon become common. It depicted the varied types of landscapes—irrigable lands, timbered lands, and timbered areas that had been burned—using different colors. This map reveals Powell's astute observation that water was a key to the future of this region, and that the landscape offered clues to its ability to support development.

By the 1860s and 1870s, many of the region's mining districts were mapped following the lead of Clarence King, George Wheeler, and others. Two maps from Nevada's booming Comstock Mining District in the vicinity of Virginia City—the *Map of the Comstock Lodes Extending down Gold Cañon* and the *Map of the Lower Comstock and Emigrant Consolidated Mining Cos. Mines* (fig. 7.7)—are instructive. They reveal another characteristic of mining district mapping—the close relationship between underground mining and above-ground settlement. Both maps were produced by G. T. Brown and Co. of San Francisco and are typical of late-nineteenth-century mining district maps. The use of pastel color washes—blue, pink, yellow—helps the maps differentiate the mining properties from one another, and also indicates the patent status of these properties. The *Comstock Lodes* map also indicates something about the geology below ground as it reveals the general trend of the "lode," which is reaffirmed by the location of active and abandoned shafts and "works." The map also reveals much about mining's relation to the hydrology: the stream in Gold Canyon provides the classic location for stamp mills, and no fewer than eight mills are indicated here. The "lode" is also depicted on the *Lower Comstock* map, as are mills and tunnels, and cultural features such as the express offices and tollhouses. The ren-

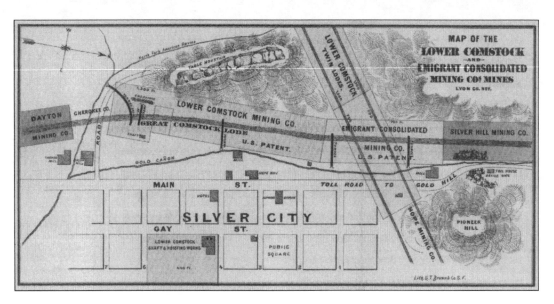

FIGURE 7.7.

G. T. Brown & Co., *Map of the Lower Comstock and Emigrant Consolidated
Mining Co[mpanie]s Mines, Lyon Co., Nev.* (ca. 1870).
Courtesy Special Collections Division, University of Texas at Arlington Libraries

dering of topography on this map exhibits some flair, as when it attempts to
depict Table Mountain's prominent geological structure and visual impact
by using bold hachure lines.

Some mining district maps, such as the 1868 *Map of White Pine District,
Lander Co. Nev Showing the Principal Mines . . . ,* are simply stunning—both
as scientific devices and as works of art. Like its counterparts elsewhere, this
map depicts towns like Hamilton and Treasure City. It also depicts White
Pine Mountain using dense hachures, offering a glimpse of the world
underground with a section through Treasure Mountain. Like a number of
its counterparts, this is a pocket map that could be folded and taken to the
mining locations in question.[23] Typical of the delineation of a developing
mining district, the 1879 *Map of Bodie Mining District Mono Co. Cal[ifornia]*
(fig. 7.8) by R. M. Smythe, M.E., depicted the locations and boundaries of
several hundred mining properties. A topographic drawing or profile was
also provided to give the map reader some concept of the terrain, and the
same prominent hills are also depicted on the map proper—which is a
planimetric delineation of the claim to the highly mineralized area just east
of the downtown section of Bodie. At this date, the Standard mine, tramway,
and mill are prominent features, and two other mills (the Syndicate and the

Bechtel) rim the district to the north. The map's private-sector publisher, the *Mining Record,* marketed a widely read journal that described and promoted mining interests. It is not known whether Smythe used a previously published survey map to produce this map, but his map was "entered according to Act of Congress in the year 1879 by A. R. Chisholm in the Office of the Librarian of Congress at Washington" and thus became part of the cartographic and historical record. Such maps owed their origin to visionary government geologists like Clarence King but flourished in the hands of entrepreneurial map publishers like Smythe, a subtle reminder that much private development has benefitted from huge public investments in information gathering.

So it was that mining left a profound mark on both the region's landscape and its cartographic history. The mapmakers used a number of techniques to depict activity both above and below ground. Some techniques, like color lithography, were state of the art and revealed an insatiable public interest in greater technical explicitness in imagery; others, like hachuring, were artistic, even beautiful, in an age that was becoming ever more scientific and technical but still valued creative cartographic flourishes. But alas, the days of the hachure were numbered. For all their implied artistry and accuracy, hachures flourished for a relatively brief period. By the 1870s and early 1880s, they were rendered obsolete by a new technique of depicting topography—the contour line. Derived from the Italian word *contorno* (to round off or turn), contour mapping uses continuous lines to delineate points of equal elevation. The term *contour map* had entered the popular language in 1862, at just the time when western exploration was revealing the lay of the land. By the late nineteenth century, the U.S. Geological Survey used contours to depict the topography, and the Great Basin was captured in increasingly elaborate detail. Like rings in a bathtub, contour lines connect points of equal elevation—in this case, above sea level. A barometer is used to help determine these lines, and further surveying and triangulation pinpoints their location. Whereas hachures were relatively impressionistic, contour lines conveyed a sense of authority because of their implied connection to transit surveying.

This last point suggests something else about contours. Although drawing them requires better and more accurate surveying, understanding them requires additional skills of the map reader—including the general map-

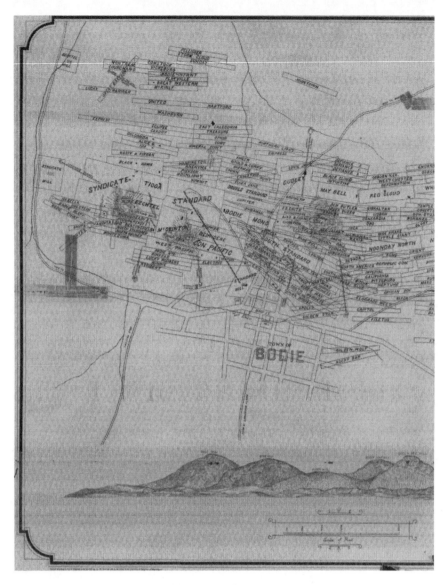

FIGURE 7.8.

R. M. Smythe, *Map of the Bodie Mining District Mono Co.* Cal[ifornia] (1879).
Courtesy Special Collections Division, University of Texas at Arlington Libraries

reading public. Whereas reading hachures is intuitive because our eye (like the hachure lines) naturally reads topography's verticality, reading contour lines to decipher the topography is counterintuitive because it requires the map reader to register on the intricacies of the form on a horizontal plane first, then extrapolate. As with isobars on a weather map, the spacing reveals the gradient: the closer the spacing, the stronger wind or steeper the slope.

MAP
BODIE MINING DISTRICT
MONO CO. CAL.
Compiled From Official Records and Surveys
By
R. H. Smythe, M.E.
PUBLISHED BY
THE MINING RECORD.
61 BROADWAY,
NEW YORK.
1879.

Contour lines suggest quantification, and it is not coincidental that they appeared on maps at exactly the same time that scientific observation came to be based on careful objective measurement. They also arrived at the same time that the general public began to think more scientifically.

George Wheeler's surveys left one additional cartographic legacy—the mapping of towns where people clustered to conduct economic activity. Wheeler's multivolume *U.S. Geographical Surveys West of the 100th Meridian* presented maps of more than a dozen important communities surveyed in 1872-73. These maps and their communities remind us of how diversified

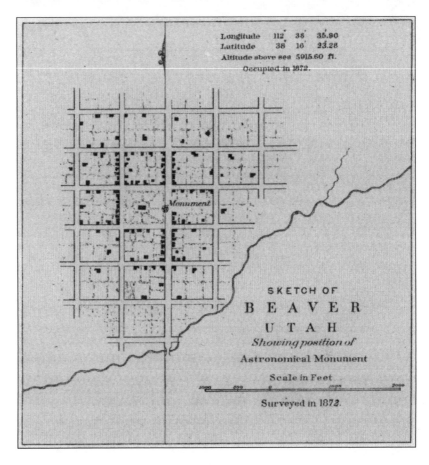

Longitude 112° 38' 35.90
Latitude 38° 16' 23.28
Altitude above sea 5915.60 ft.
Occupied in 1872.

Monument

SKETCH OF
BEAVER
UTAH
*Showing position of*
Astronomical Monument

Scale in Feet

Surveyed in 1872.

FIGURE 7.9.

George Wheeler, *Sketch of Beaver, Utah,* reveals the orthogonal
geometry of a classic Mormon village. From Wheeler, *Geographical Surveys
West of the 100th Meridian* (1873).
Courtesy Special Collections Division, University of Texas at Arlington Libraries

the region had become. They include mining towns (Pioche and Virginia
City, Nevada), railroad towns (Battle Mountain, Carlin, and Winnemucca,
Nevada), and Mormon farm villages (Beaver and Saint George, Utah) (figs.
7.9, 7.10, 7.11). Those railroad towns served diverse interests, including
ranching, and their streets were usually aligned to the rail line. Mining towns
were dedicated to unearthing treasure and serving populations bent on sep-
arating miners from their wealth, and they might be carefully planned or laid
out helter-skelter. By contrast, the Mormons discouraged mining for precious metals and focused their attention on building self-sufficiency through
agriculture. Under Brigham Young's leadership, their mines were developed
for coal, iron, and lead—not silver and gold. Their numerous farming vil-

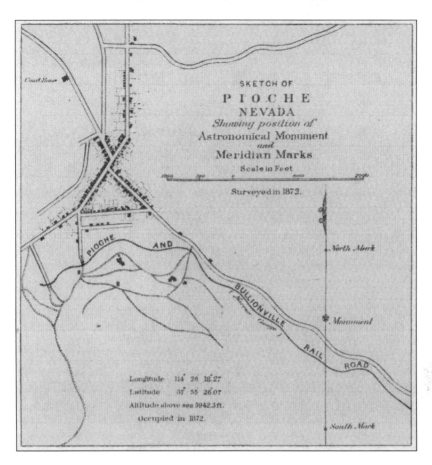

FIGURE 7.10.

George Wheeler, *Sketch of Pioche, Nevada,* typifies the irregular design
of unplanned, rapidly growing mining town in the Great Basin.
From Wheeler, *Geographical Surveys West of the 100th Meridian* (1873).
Courtesy Special Collections Division, University of Texas at Arlington Libraries

lages were almost invariably laid out in a rigid grid oriented foursquare on
the compass. Wheeler's maps depict the positions of astronomical monuments as well as buildings and property lines.[24] Although these community
maps were called "sketches," that is a bit modest. They resulted from careful
surveys and amounted to key pieces in an evolving cartographic jigsaw puzzle. The remainder of the region, though very lightly populated, was also
surveyed, and this process helped pave the way for additional mining, ranching, and agricultural activities.

Wheeler's interest in representing mining was understandable, for the
economy of the region depended on it. In addition to the hope that farming would take hold here, which was a driving force for government sur-

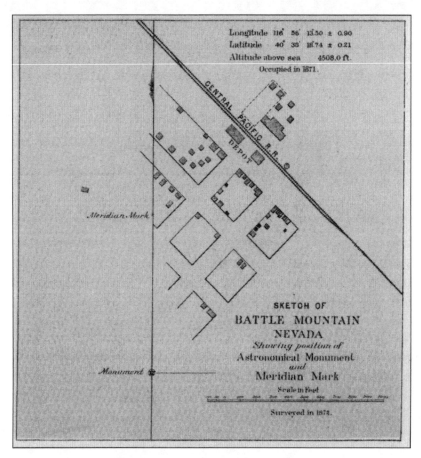

FIGURE 7.11.

George Wheeler, *Sketch of Battle Mountain, Nevada,* reveals the town's
orientation to the Central Pacific Railroad line.
From Wheeler, *Geographical Surveys West of the 100th Meridian* (1873).
Courtesy Special Collections Division, University of Texas at Arlington Libraries

veys, mining was responsible for much of the articulation of the interior
Great Basin on maps. We see the pattern set in the late nineteenth century,
when maps of this region became more and more detailed as they indicated the locations of mining districts.[25] Those mining towns helped stimulate
other activities, including ranching and farming. During the later nineteenth century, intrepid prospectors began to move south into the most formidable portion of the region—its southwestern edge, where low elevation
translated into intolerably high summer temperatures. Although the mining
town of Silver Peak blossomed by the mid 1860s, it would take another
three and a half decades—until 1900, to be exact—for miners to discover
southern Nevada's treasures. In that year, the name Tonopah galvanized the

FIGURE 7.12.

Utah as the Promised Land: map from *Pointer to Prosperity,* a brochure published
by the Rio Grande Western Railway in 1896. Author's collection

mining world as thousands flocked into southern Nevada. These prospectors were part of a concerted effort to develop the entire region, which was
now nominally gridded by a rectilinear survey system. Keyed to that system
was a series of topographical maps produced by the U.S. Geological Survey
using contour lines at a scale of 1:62,500.

By 1896, with Utah statehood achieved after the Mormons finally abandoned polygamy, the region's dual personality was formalized—bawdy, free-
wheeling Nevada to the west and sober, conservative Mormon Utah to
the east. One other north-south line on the map helped underscore this
division: With the separation of the United States into time zones in the

1880s, the line between Mountain and Pacific was ultimately drawn along that same Nevada-Utah border. The effect of this north-south split was to fragment the majestic sweep of the awesome Great Basin into two more-manageable parts. Interestingly, numerous observers equated the Mormon settlement of the region with the Jews' finding a homeland in the Holy Land: Upon their arrival here in July of 1847, the Mormons themselves had encouraged such biblical comparisons, naming the river that flows between Utah Lake and the Great Salt Lake, the Jordan. Such comparisons existed throughout the late nineteenth century, and mapping helped draw the connections. On a map in a popular brochure titled *Pointer to Prosperity,* published by the Rio Grande Western Railroad in 1896, Utah Lake is interpreted as the Sea of Galilee, while the Great Salt Lake becomes the Dead Sea (fig. 7.12). To achieve this cartographic analogy, the Holy Land's features are in effect drawn upside down (that is, south is at the top of this inset portion of the map), but as the map's title claims, the comparison is indeed striking: The Mormon's Deseret is the New World Zion—a veritable promised land—and that claim is substantiated by this ingenious map that is both promotional and inspirational.

By century's end, selected areas of the Great Basin—especially mining districts and irrigated land—were mapped. However, large poorly mapped (in any detail, at least) areas remained despite the nation's goal of mapping the *entire* Great Basin via USGS topographic maps to stimulate development and increase knowledge. Thus, by 1900, USGS topographical maps were recognized as the standard that would confirm the ultimate mapping, and   taming, of the last wilderness region in the American West. By the beginning of the twentieth century or shortly thereafter, when the mining towns of Tonopah and Goldfield were created, the region was losing its reputation as unknown. *Terra incognita* had been banished to other, even more-remote, parts of the world, namely the Arctic and Antarctic. The Great Basin was no longer a void. Rather, it had become part of the grid—a rational order that brought control to the wilderness.[26]

# 8
# Maps of the Modern/Postmodern Great Basin
## 1900–2005

THE TWENTIETH CENTURY BROUGHT phenomenal changes to the Great Basin. Several factors, including the development of the automobile and air travel, were to revolutionize both society and place here as elsewhere. Sites associated with the defense industry flourished, and cities like Las Vegas, Salt Lake, and Reno boomed. And yet many traditional activities begun in the nineteenth century, such as ranching, persisted. Mining also maintained its strong presence in the Great Basin. Consider events in the booming Bullfrog Mining District of southern Nevada. The name Bullfrog was derived from the mottled greenish ore that had been found here, and it proved irresistible as a name for the mining district. It also proved irresistible as a cartographic icon. In the early twentieth century, a humorous map delineating the Bullfrog Mining District was designed by T. G. Nicklin of the *Bullfrog Miner* newspaper in Beatty, Nevada (fig. 8.1). Nicklin's imaginative map reveals how whimsical cartographic images could be in an era of aggressive competition among towns and mining districts hoping to lure capital and people. Nicklin depicted Beatty as the literal heart of the Bullfrog Mining District, and he creatively employed the railroads (and projected railroads) as arteries in this anatomy lesson. The cartographer transformed the namesake of the mining district—a stylized bullfrog—into something that everyone could identify with. On his ingenious map, the bullfrog awaits our scrutiny much like a laboratory specimen about to be dissected by a student. Nicklin's map created quite a stir when it was published. Several observant wags noted that while Beatty was the heart, the town of Bullfrog was positioned in the armpit.

At just the time that Nicklin's map was produced (1907), southern Nevada's mining communities were thriving. The main streets of new boomtowns like Tonopah (1900) and Goldfield (1905) were crowded with people on foot, horse-drawn wagons, and that new phenomenon of the new century—automobiles. Within a decade, the Great Basin found itself on a rapidly developing national network of roads and highways. Even before the development of improved roads, however, bicycle relay racers and automobile-driving pioneers traversed the region using trails that still bore the tracks of horses, oxen, and wagons. When daredevils journeyed across the region on the first coast-to-coast automobile expeditions, they used any and all surfaces to get them through. These horseless carriages soon evolved into automobiles that had to traverse primitive, rutted roads. These vehicles were little more than internal combustion engines mounted on wagons or buckboards. On several occasions, automobiles were driven right on the railroad tracks, the drivers ever alert for oncoming trains that now barreled along the SP mainline girdling the region's midsection. These early automobile adventures not only tested the drivers' mettle, but also provided an opportunity for early automobile manufacturers—Olds, Hudson, Franklin, Packard—to demonstrate the punishment that their cars could withstand.

Lester Whitman and John S. Hammond were typical, perhaps, of the intrepid early motorists who challenged the Great Basin. In midsummer of 1903, they set out from San Francisco in a new Oldsmobile with a letter from the mayor addressed to his counterpart in New York City. As charted, their route would take them directly across the Great Basin from Reno to Ogden. In his log, Whitman humorously noted that their valise did not "contain tuxedos or stovepipe hats, but was to hold our film rolls, writing materials, our crude maps, such as they were, and the letter which we hoped to receive and deliver between the two mayors. We carried our complete wardrobe on our backs."[1] Not trusting their maps, they also carried a compass. As they prepared to leave Reno, Whitman noted that the duo "set about to fortify ourselves against the demons of sand, sun and thirst." That effort included securing "a capacious and practical sort of canteen to carry our drinking water."[2]

Fifteen days later, after numerous misadventures that included running

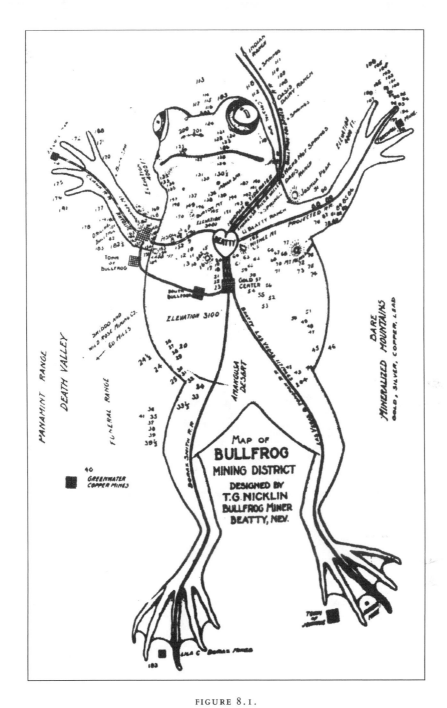

FIGURE 8.1.

T. G. Nicklin, Map of the Bullfrog Mining District (*Bullfrog Miner*, 1907).
Author's collection

out of water and several mechanical breakdowns, Whitman and Hammond had finally crossed Nevada. Fully half a day had been spent repairing the road between Elko and Deeth. Like some modern motorists in the Great Basin, Whitman and Hammond also faced another problem—locating sufficient gasoline. The Utah desert, especially the salt flats bordering the Great Salt Lake, was particularly bleak, but they made much better time here—100 miles in one day. They entered Ogden on July 25 "with cracked lips and blistered faces."[3] By mid-September, to the relief of their families and the Olds motor company, they reached the East Coast, proudly handing the letter from San Francisco's mayor to his counterpart in New York.

Whitman's statement about the "crude" maps that they carried deserves additional comment. His comment "such as they were" suggests considerable discrepancy between the maps' content and the areas that they depicted, or were supposed to depict. Given the vast expanses that Whitman and Hammond traversed, their maps served only as general guides. True, the maps contained sufficient information about the main wagon roads, railroads, and landmarks to keep Whitman and Hammond generally oriented, but the two motorists became lost enough on several occasions to resort to a time-honored tradition—asking directions from locals. In the middle of the Great Basin, where habitations were few and far between, this often involved considerable backtracking. Like mariners lost at sea, Whitman and Hammond relied on their compass to reveal the general direction of their meandering course through the sagebrush ocean (fig. 8.2). That instrument and several palpable clues—such as the sound of a locomotive whistle, the presence of a mountain range, the location of the Humboldt River, even the position of the sun—helped to keep them generally on course.

Even today, these are the types of clues that the backroads traveler must use. In June 2001, while traveling in the backcountry west of the Carson Sink, I set out on an unimproved road that was clearly marked in the *DeLorme Atlas of Nevada*. Suddenly, just beyond an irrigation ditch that also showed on the map, the road I was traveling on simply ended. Incredulous, I gazed northward through the sagebrush and shadscale, but there was no trace of a road. Yet there it continued on the map. This is a reminder that the disclaimers on the mapmakers' products—"the information in this atlas was correct to the best knowledge of the publisher at publication time, but is subject to change"—are put there for a good reason. This point, coupled

FIGURE 8.2.

Lester Whitman and new Oldsmobile at unidentified location in Nevada, 1903.
Courtesy Oldsmobile History Center, Lansing, MI

with the fact that most mapmakers build in occasional deliberate errors—a spurious place name, a road where none exists, a nonexistent island in a lake—ought to make one wary of trusting everything on a map. Those errors, by the way, are put on some maps as "signatures" or "hooks" that can help mapmakers prove their case when they take a copyright violator to court. Pity the poor plagiarizer of a map that is produced and marketed with one of these signatures. He will have to explain how in the world he depicted something that does not exist or exists in only one other place— on the map whose maker is suing for damages.

Drivers like Whitman and Hammond not only demonstrated the feasibility of the automobile; they also demanded better maps to orient themselves. At first, they used any maps they could find, including military maps showing routes, water holes, springs, and other features. However, because the actual surfaces traveled on were so critical, the automobile drivers soon demanded a new type of map—the road map—that outlined the roads and highlighted specific intersections where important decisions had to be made. Although mapmakers in distant cities began to take an interest in cross-country road maps, some of the more effective maps and guides for early motorists were developed and marketed within the Great Basin itself. "To my knowledge, we had the first tourist bureau of information in the country here in Salt Lake," Bill Rishel recalled of his career in the early 1900s as a producer of strip maps that depicted specific routes. Virginia Rishel added, "The automobile strip maps were the first to be drawn. The route book was the first ever published" in 1906. Under the headline "Pathfinder Car Maps Out Utah Auto Routes," the *Salt Lake Tribune* in 1911 rejoiced about the availability of a "Log [that] shows all turns in [the] road." Here, in narrative form, motorists would receive "reliable information which is sure to increase auto touring and to make the car of the resident and tourist a common sight in all parts of the state." Once people felt confident in finding their way, travelers would help put the state's towns "on the map" by bringing "city people out into the country." As the genius behind these logs and maps, Bill Rishel was called the "Pathfinder" and became automobile editor of the *Salt Lake Tribune*. Rishel had begun his career as a bicyclist but easily made the transition to automobile advocate.[4] The nickname Pathfinder is appropriate here, for it dates from John C. Frémont's exploits in the Great Basin more than half a century earlier.

Some of the guides offered nonvisual, which is to say essentially verbal, information. A motorist leaving Winnemucca, Nevada, might be instructed to "cross the Southern Pacific railroad tracks, then turn right (east) on first road at .10 mile; then drive 6 miles to railroad water tank; then turn right (south) and cross tracks and continue along road paralleling the Humboldt River." Knowing where to turn depended on local landmarks and the mileage to that particular location. Odometers (or trip meters) actually predated the automobile; the Romans invented the first, and wagons used them throughout the nineteenth century.[5] With guide in hand, motorists could navigate more or less effectively, but they often carried road maps as well. These depicted larger areas and enabled travelers to comprehend alternative routes. Because the states oversaw the developing networks of roads, maps broke the region into several jurisdictions. However, with the development of the national highway system after 1921, people began to think in terms of subsidized federal highways, many bearing names like Lincoln Highway. These state and federal routes, combined with a series of private roads upon which tolls were charged, began to form an integrated system by the mid-1920s. As service stations proliferated, oil companies began to offer road maps as incentives to patronize their facilities.

When commercial road maps became widespread, it wasn't long before motorists began modifying them to indicate important landmarks. Thus, if locals advised someone to turn at a railroad water tank six miles from town, a motorist might pencil in a 6 and mark the location of the tank with an arrow. Thus it is that even commercial printed maps may be supplemented by the map reader—who bridges, in an informal but time-honored tradition, the gap between mapmaker and map user.[6] By the 1930s, motorists would not think of traveling far without road maps. With them in hand, the Great Basin was no longer a mystery to the average traveler. Road maps opened up options not available to those confined to rail travel. Whereas the passenger on a train could trace his or her progress on a railroad company map, he or she was confined to that route. By contrast, the automobile driver now had an increasing number of choices. Moreover, scenic locations and other points of interest could be reached by car, and road maps often depicted these.

The history of road maps is worth recapping. Studied through time, they reveal the increasing tendency for the region's geography to be articulated visually in popular culture. In a fifty-year period, ca. 1900-1950, the road map underwent a steady evolution as motorists became more literate, graphically speaking. Motorists first relied on guidebooks such as those by the Automobile Club of America (founded in 1899) and the American Automobile Association (1902). These publications contained narrative descriptions, but road maps produced by oil and tire companies began to proliferate after 1905. By the 1910s, road maps included portions of the Great Basin. Rand McNally's 1921 road map depicted a growing network of roads and highways across the region, as did the George F. Cram Company's 1922 *Auto Trails and Commercial Survey of the United States.* Even the formidable Great Salt Lake Desert was now traversed by two routes that appear as bright red lines slashing across this once fearsome landmark of salt flats. The Great Salt Lake Desert's depiction as a flat sheet stippled to indicate its parched, powdery surface surrounded by rugged, hachured mountains (fig. 8.3) enabled the motorist to navigate using the tools of the explorer. The late 1920s witnessed the flowering of the oil company road map, which became an American art form—compact, informative, attractive, and unabashedly commercial. By the 1940s and 1950s, sectional maps of the western United States were complemented by large-scale route maps prepared by the AAA.[7] The finest road maps of the 1940s and 1950s depicted not only the roads in detail (density or color indicated width, number of lanes, and type of surface), but also the types of communities (their size and function—as, say, a county seat—indicated by the size and color of dot) encountered. This was information motorists needed to know, because they were dependent on the services—auto repair, accommodations, food—found there.

By 1955, when Frank Sinatra sang a famous line from the title song in the romantic movie *The Tender Trap*—"You hurry to a spot that's just a dot on the map"—everyone who heard it could visualize the hierarchy of places. Aside from Salt Lake City, Reno, and the upstart Sunbelt gambling mecca of Las Vegas, every community in the Great Basin was just a dot on the map. Road maps did more than inform; they also helped shape perceptions. By breaking the region into state-shaped pieces, road maps further helped fragment the Great Basin in popular thought. Although the region was found on maps of the western United States and represented in one sweep on AAA TripTik maps, these were relatively rare compared with the free state road maps that oil companies handed customers who filled up at their service stations. In the 1950s and 1960s, most map users consulted separate road maps—Utah, Nevada, Oregon—to find their way through the region. (They still do, but at a price: Although service stations no longer give out free road maps, one can still purchase them there or at the many speedy marts selling gasoline along with snacks, drinks, and personal items.) Even that wonderful travel accessory of the late twentieth century, the road atlas, fragments the region state by state. Furthermore, each state is mapped in detail that reveals a complex mosaic of land ownership and political jurisdictions at the local level. That political jurisdictions affect and reflect our sense of reality is evident in the myriad of jurisdictional items—counties, Indian reservations, land administered by dozens of federal agencies—that now configure maps of the region into collages of varied sizes, shapes, and colors. By doing so, they ultimately shape our perceptions, further eroding the concept of the Great Basin as one unified region.

Road maps of the Great Basin helped demystify the region at the very time that improvements in highway design and automobile technology made travel safer and faster. A billboard photographed near Tonopah (fig. 8.4) in 2000 suggests how seamlessly road maps have become part of our perceptions. The designer of the billboard knows that the motorist who has reached Tonopah still has a long way to go before anything "exciting" hap-

FIGURE 8.3.

Portion of Utah road map by George F. Cram (1922) shows the formidable
Great Salt Lake Desert traversed by roads.
Courtesy Special Collections Division, University of Texas at Arlington Libraries

FIGURE 8.4.

Billboard at north edge of Tonopah, Nevada, creatively incorporates a stylized
road map. June 2000 photo by author

pens—hence the placement of a casino as a destination at the road's end.
To promote the casino, the designer employs the most basic elements of
mapmaking—a destination marked at the end of a stylized line of travel.
This billboard is, in fact, a highly abstracted road map that exploits an interesting technique: The position of the highway symbol and the use of perspective subliminally place the viewer in the lower left corner of the composition. The motorist thus knows that he or she is headed toward the
destination, which is made all the more irresistible in the context of the surrounding isolation of the desert. This billboard as stylized map reminds us
how much we have come to identify with, use, and trust maps. It also confirms the now easy alliance between cartography and advertising.

Consider another billboard-as-map, located at the junction of U.S.
Highway 93 and Nevada Highways 375 and 318 (fig. 8.5). This one is much
more explicit and less abstract but still serves the same purpose as the billboard near Tonopah. The motorist who experiences it likely has a road map
by his or her side but now encounters a huge, easily recognizable map as a
feature in the wide-open Great Basin landscape. This is not an objective
sign/map, but rather another example of a map/sign as behavior influencer.
Encountering it at a crucial decision point about 100 miles northeast of Las
Vegas, motorists are in effect persuaded—that is, lured—to steer to the right,
or eastward, toward Caliente and several scenic parks. The sign's stylized
mountains promise a more scenic, and hence rewarding, drive in the age of

FIGURE 8.5.

Billboard at highway junction in Nevada uses road map to persuade the traveler to
take one route over another. April 2003 photo by author

amenity-based tourism. The sign's main message—turn right here—is made
credible by that device which always conveys authority, a map. This billboard-as-map is a subtle reminder that maps work to inform us about the
growing number of options we have, including the ever-present business
interests who hope we will select their products and services. This sign is
also a reminder of how many bewildering choices the motorist faces compared with his or her counterpart on a train.

The traveler on a train rushing across the Great Basin in the early-to-
mid-twentieth century was likely to have purchased a ticket from one of the
major regional railroads—Union Pacific, Western Pacific, or Southern
Pacific—and was often given a timetable that featured a map. In the case of
the Southern Pacific, the route across the region was part of the Salt Lake
Division. The railroad's official map of this division dated August 1919 features a wealth of information about the region's hydrology, topography, and
deserts, as well as cultural/political jurisdictions like counties, national
forests, and Indian reservations (fig. 8.6). The maps given to travelers were
modified to fit into the format of a folding timetable crammed with information about the route. In addition to showing the mileages from station to
station and the times at which the trains were scheduled to arrive and
depart, the maps of the route showed the railroad's line in relation to features of interest in the adjacent countryside.

The Southern Pacific's route through the Great Basin was called the

"Ogden Route" until about the mid-1920s, when the name "Overland Route" was adopted. Southern Pacific timetables from the 1930s to the 1960s opened accordionlike to reveal the railroad's route and the prominent features along it. This type of timetable/map served two purposes. With it, railroad travelers could keep track of the train's progress across the region and also become better informed about the scenery along the route. By the 1950s, the Southern Pacific's Overland Route promotional materials featured stylized maps meant to suggest an interesting and rapid trip (fig. 8.7). Tellingly, on this simplified, cartoonish map, only a few points of interest were found in the Great Basin, the mining town of Virginia City, the majestic Great Salt Lake, and a lone horse and rider being the exceptions. This was still wide-open country, and the railroad's map confirmed it. Similar maps were employed by the Western Pacific, which pretty much paralleled the Southern Pacific across the region's midsection, and the Union Pacific, which traversed the eastern and southern edge of the region on the Salt Lake City-Las Vegas-Los Angeles route.

The railroads also employed maps to lure shippers in addition to passengers. In 1971, at the dawn of Amtrak, the Union Pacific prepared a folding "Geographically Correct" *Map of the United States* (fig. 8.8) that showed their lines in bold red—with the lines of other railroads thinner and much less conspicuous. The Great Basin portion of that map is especially instructive, for it stands out from the rest of the mapped area as a largely white or blank space. To further distinguish its railroad line traversing the eastern edge of the region, the Union Pacific used a technique that had been employed as early as the late 1880s to show virtually every location, even small tank towns, as named dots. Rand McNally had apparently pioneered this technique as the company became the premier producer of railroad maps by the late 1880s. Cartographic historian Kit Goodwin calls these "designed maps" because they are carefully engineered to suggest that the railroad's line traverses a populated and well-settled area with evenly spaced towns.[8] Those dots along the railroad, like so many beads on a chain, further contrast with the emptiness of the Great Basin. A map of this type reveals the close connection between the railroads and settlements. Where no railroads exist, as in a huge section of Nevada, a few scattered dots like Goldfield and Currant underscore the area's isolation, while a stippled pattern in western Utah indicates the formidable Great Salt Lake Desert.

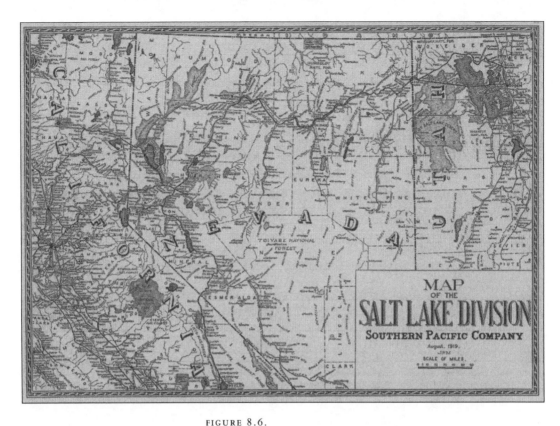

FIGURE 8.6.

Map of the Salt Lake Division, Southern Pacific Company (1919). Author's collection

An even more stylized map by the Western Pacific Railroad reveals how seamlessly cartography and advertising worked together by the mid-twentieth century. Combining a simple map of the route of the California Zephyr with a line drawing of a diesel-powered train (fig. 8.9), the advertisement's horizontal sweep reaffirms the Western Pacific's main purpose—to move people and goods across the Great Basin efficiently. The map and train are rendered with a splash of the orange coloring that the Western Pacific integrated into its paint scheme—another reminder of how smoothly, and subliminally, a map could reinforce corporate identity.

In order to keep trains running smoothly and safely, railroads also produced maps for their employees. These maps were rarely or never seen by the public, but they contained additional detail needed by workers, including the number of tracks at particular locations and trackage agreements with other railroads. The Western Pacific Railroad's system map (fig. 8.10) shows the route from Salt Lake City to Reno Junction as either a double-

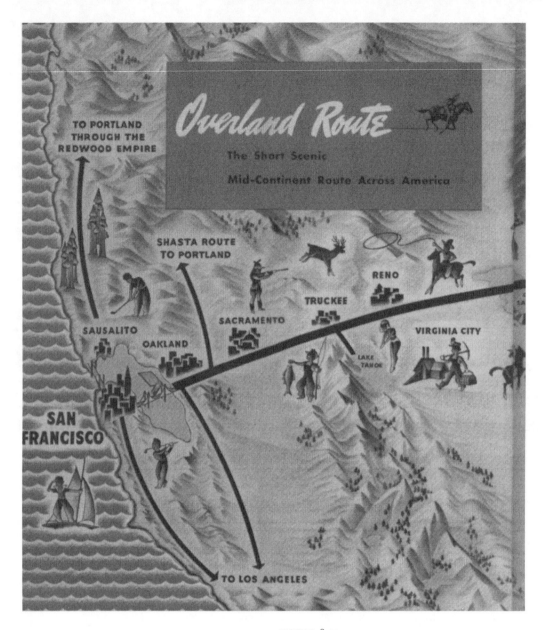

FIGURE 8.7.

Southern Pacific pictorial map(s) of the Overland Route
showing Great Basin (ca. 1955).
Author's collection

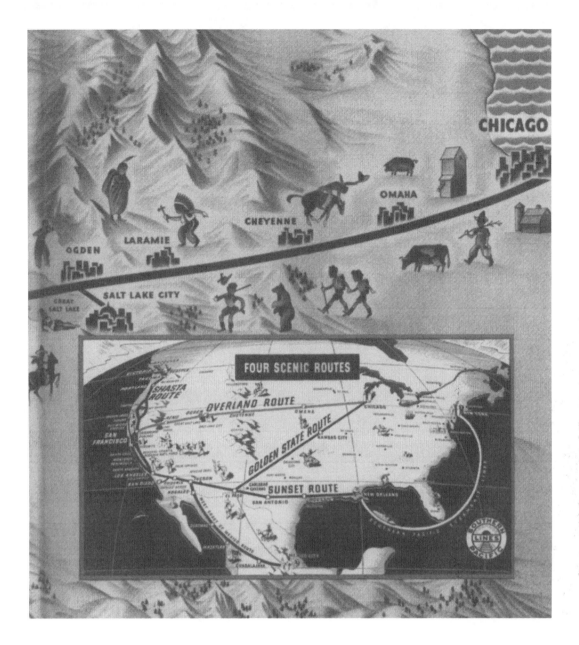

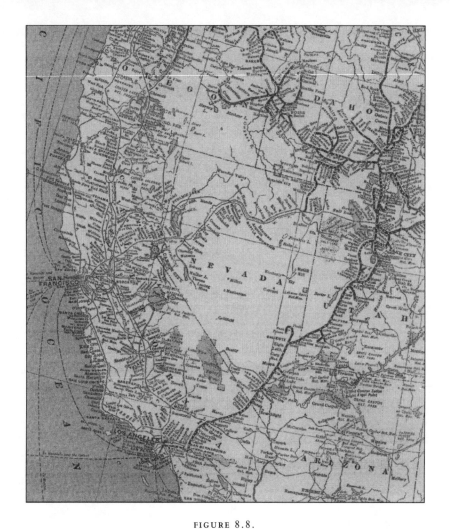

FIGURE 8.8.

Detail from Union Pacific Railroad's "Geographically Correct"
*Map of the United States* (1971). Author's collection

track mainline or operated under Centralized Traffic Control (CTC). It also
shows the location of the Western Pacific's paired track operation with the
Southern Pacific. Even on this technically oriented map, something of the
openness of the Great Basin can be grasped by the large amount of blank
space on both sides of the railroad line. The fact that CTC was controlled
by dispatchers in distant cities who consulted wall-mounted illuminated
maps of the railroad and its many signals further underscored the isolation
of railroad lines in the Intermountain West.

Centralized Traffic Control improved the railroads' speed and increased
safety, but highways continued to take away their revenues. At about the

FIGURE 8.9.

Western Pacific Railroad advertisement featuring train and map (ca. 1955).
Author's collection

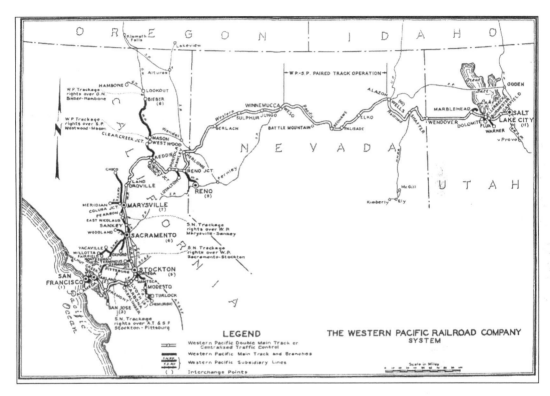

FIGURE 8.10.

Western Pacific Railroad Employee's Map (ca. 1955). Author's collection

same time that the automobile was surpassing the train, however, travelers
also began to fly from destination to destination. Air transportation has a
venerable history in the Great Basin. It, too, developed as part of a national
transportation network. By the early 1920s, just when the national highway
system was taking shape, the United States Postal Service pioneered a series
of transcontinental air routes that greatly reduced the time mail took to
reach the West Coast. Like the highway system, and to a lesser extent the

rail system, air transportation was heavily subsidized by the federal government. Begun in 1923, the system was operational in 1927. One route traversed the Great Basin from Salt Lake City to Sacramento and thence to the San Francisco Bay Area. This air route was in direct competition with the transcontinental railroad routes (Southern Pacific and Western Pacific), but because airmail rates were higher, the railroads still had plenty of regular mail business.

Safe and effective air transportation depended not only on improvements in airplane technology, but also on improved navigation and mapmaking. The U.S. Postal Service's lighted airway routes across the region signaled a revolution. Beacons and radio signals to guide aircraft were marked on maps called aeronautical charts. Given the limited ranges of aircraft at this time, the planes stopped in several locations where budding airstrips developed. Pilots depended on a wealth of information and technology but still relied on spotting features on the land such as mountains, towns, and railroad lines. No form of transportation is more dependent on exact geographic location, and the aeronautical charts featured readings of latitude/longitude as well as routes designed to ensure that airplanes remained on course and avoided each other.

The Civil Aeronautical Act of 1938 strongly supported air travel for not only mail but also passengers. The railroads had begun to add streamlined trains in the 1930s, but they were no match for air travel. By the late 1930s, regularly scheduled airliners crossed the Great Basin on fixed intercity routes. The airlines used maps showing routes and major landmarks to help orient passengers and put them at ease. In 1939, TWA (Transcontinental & Western Air) familiarized passengers with its new *Airway Map and Log*. The text reassuringly informed readers that the airplane followed "INVISIBLE HIGHWAYS" and employed numerous technological devices—radio range receivers, radio direction finders, radio marker receivers, and aircraft radio telephone equipment. To further reassure passengers, a map was included (fig. 8.11). "AIR MAPS ARE LIKE ROAD MAPS," the log stated. "Map experts, cooperating with TWA's Navigation Department, have taken great care to make these maps helpful and accurate in every detail." Using these maps, the airline noted, "[p]oints of interest can be easily identified, helping passengers orient themselves and pass the time pleasantly."[9]

The passengers on TWA flights unfolded the map to reveal the American

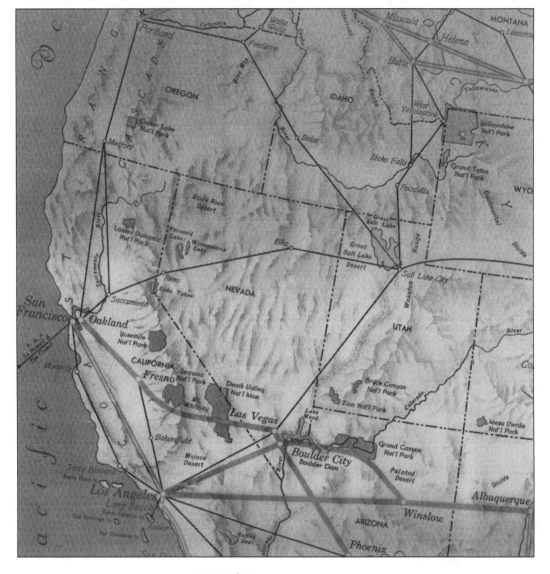

FIGURE 8.11.

*TWA Airway Map and Log,* Transcontinental & Western Air, Route of the Sky Chief
(1939). Courtesy University of Texas at Dallas, History of Aviation Collection

West crisscrossed by a series of straight and gently curving lines indicating
air routes. "The Sunny Santa Fe Trail" is emphasized on this map and log as
TWA encouraged travel on its routes in the Southwest. TWA flights from
Phoenix to California traversed the southern edge of the Great Basin, with
Boulder City and Las Vegas as major points on the route. The map shows
the remainder of the Great Basin as largely open or empty space, but a route
of another airline, most likely United, also crosses it from Salt Lake City via

FIGURE 8.12.

United Airlines, *Maps of the Main Line Airway* showing Great Basin section
(ca. 1942). Courtesy University of Texas at Dallas, History of Aviation Collection

Elko to Reno and then on to San Francisco and Oakland. The topography
is indicated by a technique pioneered earlier in the twentieth century—
shading. This technique made the topography stand out as if seen from an
airplane, a perfect complement to air travel. It makes a flat map appear to be
a relief map and is achieved by showing the topography as if it were illuminated from the northwest, throwing the east and south sides of mountains
into shadow. On this map, the Great Basin's generally north-south-trending
mountains stand out boldly in the otherwise largely blank space.

By the early 1940s, United Airlines offered comprehensive service in the
West, operating across the Great Basin on routes from Salt Lake City to Los
Angeles and Salt Lake City to San Francisco. United gave its passengers a
brochure entitled *Maps of the Main Line Airway* (ca. 1942) illustrating its
routes in considerable detail (fig. 8.12). For example, the Salt Lake City-San
Francisco route map shows the flight crossing the Great Salt Lake Desert,
slightly changing its course at Elko. It then heads toward Reno, passing over
the Humboldt River and Carson Sink, which are well marked on the map.
Along the route, places are described in text just below the map. By reading this information, passengers learn that Beowawe, Nevada, is "notable for
tall poplars" and that the name "means 'gate' in Indian language, and was so
named for characteristic formation of hills." Of Carlin, Nevada, the map
guide urges the traveler to "Note large white letter 'C' on hillside"—a
reminder that the propeller-driven airliners still flew low enough for pas-

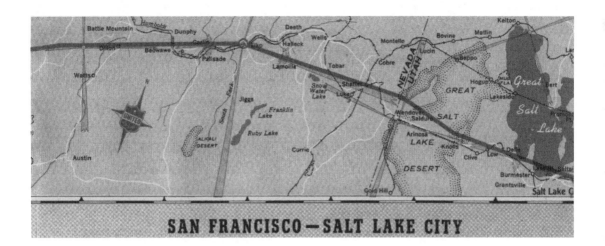

SAN FRANCISCO—SALT LAKE CITY

sengers to make out relatively small landmarks that would be indistinguish-
able from today's high-flying jets. The map contained several aerial photo
images of sights along the way, such as the Great Salt Lake. Interestingly,
however, one of the aerial photos is completely blocked out, the word "cen-
sored" across where it would normally be—a not-so-subtle reminder of
national security concerns in wartime.[10]

This same United Airlines map/guide notes that the flight traverses a historic route of the pioneers: "For a century . . . The Main Line of Coast-to-
Coast Transport and Communication." Building on the theme of the West's
romanticized past, the guide states that the "Overland Trail is a name written boldly across many pages in the history of our country . . . thrilling
exploration, hardy pioneering, bitter conquest, empire building." Today,
the guide boasts, this "short, strategic route" that once witnessed "prairie
schooners, the famous Pony Express . . . the first transcontinental telegraph
. . . the first Iron Horse to link East and West" has become "the central year
'round air route for the expedited transportation of passengers, mail and
express" by United Airlines.[11]

By the 1950s, when railroads added one final spate of streamliners in a
last-ditch effort to compete with air and highway traffic, the die had been
cast. The airline passenger traversed the Great Basin in a fraction of the time
that it took road or rail travelers—and then air travel became even swifter
after 1958 with the introduction of jet airliners. By the early 1960s,
American Airlines—"Route of the Astrojets"—offered its passengers a beautiful shaded relief *System Map* (fig. 8.13). It reveals the increasing level of

detail that the public could comprehend as the topography is rendered both artistically and scientifically. The Great Basin stands out in lighter shading, its north-south-trending mountain ranges boldly depicted in considerable detail. The Humboldt River and numerous small towns (Austin, Eureka, Wells) are also shown, as is the unnamed Southern Pacific railroad route. The American Airlines route from New York/Chicago to San Francisco is shown in red.

This map is as much science lesson as promotional tool. Flipping it over, travelers beheld a colorful verso revealing the "New Dimension Below"— a geological interpretation of the country seen passing beneath the airplane. The Great Basin is shown as an area of contorted rock layers. The airline passenger is urged to understand the "Land Below. Pause for a moment, if you will, and look at the land below. Watch for a while the green hills, and snow-capped mountains, the lakes and rivers, the great plains stretching to infinity." The text then adds an informative, even introspective, statement: "A map is more than a representation of the surface of the earth, a guide from place to place. It is a record of history, a spur to memory, a fascinating tale of brave people and brave deeds, a key to beauty."[12] But this map is more than a tribute to aesthetics and patriotism. Issued at the height of the Cold War, when the United States had grave concerns about falling behind in the space race after the launch of the Russian spacecraft *Sputnik* in 1957, it reveals a desire to educate the public in science. This map, in fact, is one of the most scientific ever offered to the traveling public up to that time. It rivals the information contained in the earlier voluminous government reports, albeit in condensed form.

Airline maps do more than depict the proliferation of companies and routes crisscrossing the Great Basin from the 1930s to the 1970s. They reveal two conflicting developments—the increasing disconnect from individual places with increasing altitude, countered by the enduring interest in understanding those same places that time and speed have come close to banishing from consciousness. Whereas the earliest maps showed trails meandering across the region from water hole to water hole, and road maps revealed a web of thoroughfares that sometimes zigged, zagged, and jogged at crazy angles, airline maps were fundamentally different in that they connected those urban dots effortlessly by straight lines. Then, with increasing accuracy and sophistication, airline maps began to depict their routes as a series of

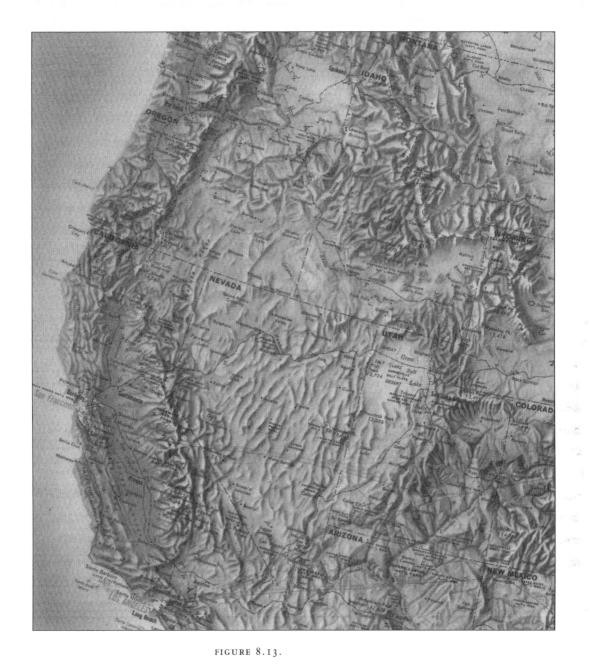

FIGURE 8.13.

*American Airlines Route of the Astrojets System Map*
(Rand McNally ca. 1961)
showing routes and detailed topography in the Great Basin.
Courtesy University of Texas at Dallas, History of Aviation Collection

gently curving arcs. Seen from the ground those planes' telltale vapor trails were razor-straight. Readers of these maps subliminally recognized that a straight line between two points curves on a map because the earth's curvature is being projected on a flat plane.

Aerial-view maps like those we have been discussing are based on a twentieth-century tradition of aeronautics. At the same time that air travelers began to have greater access to the Great Basin, gazing down upon it from a mile or two, then four, five, six, seven, even eight miles high, aggressive air photography supplemented detailed topographic mapping. Some of these photos were produced as stereo pairs that provided the viewer an eye-popping three-dimensional image. As this increased visual access developed, however, another factor was also playing out in just the opposite direction. Since before the Second World War, the Great Basin had become a major venue for military testing, some of it highly secret. We have seen that some locations were already off limits to air travelers, visually speaking, but such military efforts sometimes altered the very landscape itself. As weapons testing escalated, some of the region's fascinating topographic features, though sacred to Native Americans, were bombed into oblivion—a task made easier by maps of the gunnery ranges. Aviators' maps, intended only for experienced pilots, offer a bewildering array of symbols and signs. They reveal that access to portions of the region was now completely prohibited as military aircraft and ordnance testing became important activities in the Cold War (1947-1989). Government secrecy was enforced as fears of espionage mounted. The Cold War fueled both paranoia and beliefs in the paranormal. It coincided with the UFO phenomenon, and the region began to develop a reputation as a place of mysterious extraterrestrial visits and even more mysterious cover-ups.

All of this political and military intrigue is evident in the modern-day maps of the region. Between narrow ribbons of highway draped from town to town lie huge areas prohibited to all but military officials with top-secret clearance. On a map, fully 86 percent of Nevada, as well as most of western Utah and the huge desert triangle of southern California, is a quilt of federal lands ringing small enclaves of private landholdings. This federal quilt is further divided into public lands where users may graze cattle or mine minerals or visit places of scenic interest, such as national parks. There are also huge weapons-testing areas clearly marked as dangerous and off limits.

Because those places are just the opposite of accessible—which is to say forbidden—they are most mysterious and enticing. As the map user's eye scans the Great Basin from east to west, he or she finds the Tooele chemical weapons base, the Dugway Proving Ground, and other areas set aside for war and its tools in Utah and westward through Nevada toward the Naval Air Station in Fallon and southward toward Yucca Mountain, where one's eye stops—Area 51.

Named with disarming honesty for its mundane position in grid number 51 of the vast Nevada Test Site, this is the most secret place in the entire region. It is where, to quote one of the many websites devoted to it, "mysterious lights began to appear" in the 1980s and "armed guards known as cammo dudes started patrolling the border." The government's alleged desire to hide both spacecraft and their alien occupants led to an obsession with Area 51, which is now synonymous with secrecy and its counterpoint—a cult following of conspiracy buffs dedicated to disclosing the truth, or their version of it. The place is reportedly so secret that the "government closed many viewpoints from which the Area 51 base could be seen."[13] Flying over it is strictly prohibited, and gaining access on foot is not only unwise but flatly unthinkable without government clearance. How, then, can one experience Area 51? The answer is the modern equivalent of exploration, a website featuring a profusion of satellite images that are, in effect, maps. This website contains several dozen such electronically viewable maps, which are actually aerial photographs keyed to a base map. Like most websites, this one is fluid—updated from time to time. Each image is dated, and the number of "hits" (i.e., visitors to each image) is recorded.

Of special interest is another area identified on this same website—Area 19. The website asks, "What is it? Some people think that it is just a peice [sic] of land owned by NTS, while some think that it is the 'real' Area 51. These are just the two extremes, most people have their opinion somewhere between these two." The website goes on to explain, "The only thing that we know for a fact is that Area 19 is really just a peice [sic] of land in the middle of nowhere." Why has Area 19 drawn such interest? According to the website, "What makes it noticable [sic] is just there's a 34.5 kilovolt power line which just stops in the middle of nowhere." The writer closes rhetorically: "I think that this is a little bit unusual, don't you?" The site notes that reporters have gained access to many portions of Area 51 but have been denied access to Area 19, heightening the sense of mystery and intrigue. Scanning this website for its intriguing aerial images of portions of the Great Basin reminds one how much maps have changed over the last century. They began as lines on paper but wound up as virtual images conveyed on screen. Even the road map is now likely to be a GPS-coordinated image that enables the automobile driver to determine his or her location with precision on a digital monitor located in the automobile's dashboard. Such GPS information reveals not only much about our location on earth, but also about our increasing dependence on orbiting satellites that are now the third point in a complicated system of triangulation that began with the government surveys of the 1850s and 1860s.

Maps of the Great Basin continue to stimulate the imagination in the early twenty-first century. In the science-fiction television series *Tremors* (2003), townspeople in Perfection Valley, Nevada, use an electronic map on a computer monitor to track attacks by a giant subterranean wormlike creature (El Blanco). El Blanco is not only dangerous but is also listed as an endangered species. Protecting El Blanco will mean that the habitat is also protected, but the creature is downright deadly when provoked. When seismic activity increases as El Blanco goes on the rampage, the computer monitor map image immediately pinpoints the location via an illuminated topographical map. The area is gridded to base maps, and those maps depict the movement of El Blanco via underground sensors. This system is the brainchild of Burt (Michael Gross), an eccentric, witty, right-wing paramilitary survivalist who wryly comments on life in the desert town that uses El Blanco as a lure to the tourist trade. *Tremors* is delightfully postmodern entertainment, part serious science fiction and part spoof on the paradoxes and absurdities of modern life. In it, the map as computer screen confirms how seamlessly technology and mapmaking inform popular culture.[14]

The first episode of *Tremors* broke new ground, artistically speaking, but the series did not stop there. In the second episode, Burt uses the archived data to retrace El Blanco's movements via the electronic map. Using the skills of a cartographic historian, Burt discovers that the federal government had created an underground biotech lab in the area. When determining where this lab's genetically modified creature will next strike (thus upsetting El Blanco), Burt and others use a more traditional topographic map. Here we see two very different types of maps side by side. In that same episode, arriving tourists also consult a more traditional map in the *Nevada Guide*. The two types of maps in *Tremors*—one high-tech electronic, the other old-style paper—provide a perfect counterpoint.[15] The *Tremors* shows subliminally confirm that maps are used not only for orientation and problem solving, but also for entertainment. They also creatively juxtapose old and new technology in mapmaking as a way of suggesting that both can be used in solving mysteries.

Consider the technological changes in just the last forty years that have revolutionized mapmaking and map reading. Mapping of the Great Basin in exquisite detail took a giant leap following the successful launch of orbiting satellites after 1960. From their perch more than a hundred miles aloft, these satellites commanded views of the entire region. Until those first space flights, no human being had ever comprehended the Great Basin in one majestic sweep. Whereas Frémont may have dreamt about it, satellites now made this single view possible. Consider the depiction of the Great Basin in the *Satellite World Atlas,* which provides "[t]wo stunning views of our world—satellite images [and] detailed topographic maps" fig. 8.14). This atlas's satellite-inspired map is static in the printed form—that is, pretty much a view taken from a stationary point high above the earth. However, by visiting the publisher's website (www.worldsat.ca), one can click onto a map that is interactive: by further clicking on the map, one can reorient oneself to it from any perspective—which is to say the viewer can effectively fly over the topography, sweeping lower or higher, or changing direction. The topography stands out in bold, raised relief, and so the feeling generated is much like being above the landscape—albeit abstracted.

This apollonian view coincided with the simultaneous development of high-resolution cameras and new films that could capture subtleties in the region's geology, vegetation, and land use. That technology ultimately privileged the public to see what only the imagination once beheld. This is more than simply a privileged view from on high. Rather, it is a kinetic experience by which the map user navigates over the place much as a fighter pilot does. The military analogy is apt, for such computerized mapping is widely used in both the virtual and the real world. Aerial surveillance and tactical air support/bombardment are its military applications, but also enable military-style computer games to serve as popular entertainment. Tellingly, the prototype that is used in the real world with deadly precision is mastered by people who learn their skills in the virtual world first. Linked to the increasingly sophisticated espionage of the Cold War and the War on Terror, satellite photography radically altered the way humankind would comprehend this region and the rest of the world.

Remote sensing involves two revolutionary concepts. As the phrase suggests, the camera can be located at great distances, but even more significant is the fact that no human being need be present to do the photography. Moreover, photography itself is dramatically augmented, for light and shadow are no longer the only images recorded. Heat (infrared) and sound (radar/sonar) were added to the array of signatures that satellites captured. Generations of satellites offered not only broad views, but incredibly detailed coverage of areas. LANDSAT imagery—that is, images taken from satellites to systematically record each parcel in a huge mosaic—now made every place on earth more accessible to the public. The Department of the Interior conceived the idea of a civilian earth-resources satellite in the mid-1960s. The National Aeronautics and Space Administration (NASA) developed and launched the first satellite to meet civilian needs. In the early 1970s, the USGS partnered with NASA to archive and distribute LANDSAT data. A number of ERTS (Earth Resources Technology Satellites) were launched and relayed data under the aegis of NASA, but by the early 1980s the LANDSAT system was transferred to the National Oceanic and Atmospheric Administration. LANDSAT was commercialized, although Earth Resources Observation Systems Data Center remained as the overseer of government archive of LANDSAT data. In 1992, Congress assigned responsibility of archiving LANDSAT data to the Department of the Interior. The USGS EROS (Earth Resources Observation System) continues to manage the LANDSAT data archive. Although the acronym EROS suggests the sensuality of passionate love and therefore seems a misnomer, the program *does* enable the inquiring lover of the earth an intimacy of sorts. This program has proven a boon to land-use managers and resource users. As data is received and archived from orbiting satellites, it provides the region's ranching, mining, agricultural, and other interests with a database that permits comparisons through time; this in turn serves as a way of monitoring the changing nature of the environment.

Satellite technology is now a part of both exploration and mapping. By the late twentieth century, maps and geographic information systems, GIS

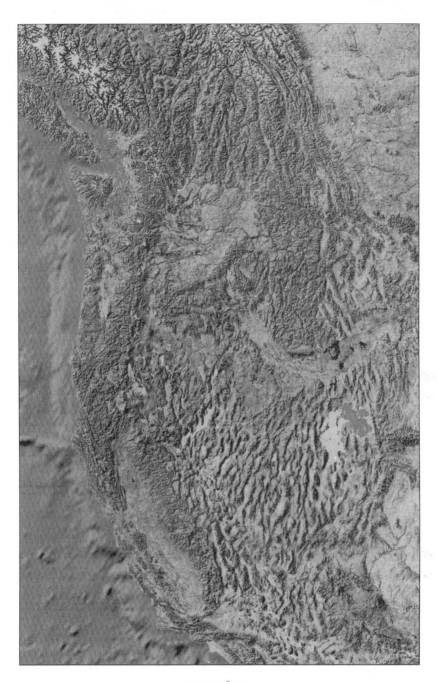

FIGURE 8.14.

Map of the Great Basin derived from satellite imagery.
*Satellite World Atlas,* 1999; reproduced with permission

for short, became almost synonymous. GIS enables varied data to be collected and interpreted for quadrants of about a square meter—or smaller. This information is veneered, as it were, using computers. GIS has demystified the earth, enabling comparisons of physical and cultural parameters. It once again confirms the age-old relationship between mapping and technology. GIS permits us to understand places in ways unheard of a century ago. Geographers pioneered GIS, but now workers in other fields—geology, biology, hydrology—use it routinely. This means that GIS has redefined and democratized geography, as many practitioners have access to a powerful tool that was once in the hands of a single profession. That is exciting, even liberating, but there is a caveat. Geographers have a rich tradition of philosophical inquiry about the spatial representation of data. Those who proceed without this knowledge are sometimes destined to reinvent the wheel—and reevaluate their assumptions about distributions of phenomena.

Overall, however, GIS has helped to revolutionize professional fields that, in turn, enrich geography as a discipline. Just as a growing understanding of complexity and diversity characterized the late twentieth century, it also affected not only disciplines and professions, but maps. Consider the physical environment that maps reveal in ever-greater detail. The twentieth century witnessed increasing interest in, and legislation protecting, the environment. The growth of the new science of ecology in the 1930s, and the environmental movement in the 1960s, led to an appreciation and protection of the physical environment, and maps were essential tools in this process. This was certainly the case in the Great Basin. Whereas the environment of this region had seemed limitless and expendable in the nineteenth century, in the twentieth century it came to be viewed as fragile and irreplaceable—at least by more enlightened policy makers and land users. Maps were essential to a growing appreciation and understanding of the Great Basin as a unique habitat. At the same time, the region's ethnography also became an important subject of study. Few places better reveal the close connection between habitat and lifestyle. As ecologists and ethnographers shared their findings, the understanding of how native peoples existed in a delicate balance with the environment grew.

But all this intensive mapping and map use did more than enhance people's understanding; it fundamentally reshaped the *way* that we comprehend and understand the region. Topographic maps and satellite images became essential tools for a wide range of people—ranchers, miners, conservation-ists—who live and work in the region. Above all, a refined understanding of the physical environment revealed that there is not one Great Basin; rather, the region's geography varies from place to place. This awareness has enabled ecologists to show that the eastern and western parts of the region vary significantly; the eastern portion is ecologically more akin to the Rocky Mountains, while the western portion is more closely related to the ecology of the Sierra Nevada and the Pacific Coast ranges. Moreover, the vertical zonation—which is to say elevation is an important factor—underscores the region's topography. Most significant, perhaps, is the realization—as William Bell suggested as early as 1867—that the Great Basin is much more complex than its disarmingly simple name implies. It consists of not one, but more than one hundred, basins of interior drainage. This suggests a new, ecologically oriented definition for the entire region. Rather than a single Great Basin, it is in fact the region of many smaller basins. That in turn links it to its even larger physiographic province—the Basin and Range region of western North America. When the many other environmental factors—temperature, precipitation, humidity, vegetation, fauna—are added to the mix, the region appears as a complex mosaic. This tapestry may look like one huge composition from afar, but closer inspection shows that it is woven from varying threads; no two parts are identical, and this fact immediately raises the specter of a myriad of unique habitats that require better understanding and appreciation.

Satellite imagery goes only so far in helping to decipher the character of this region. In the twenty-first century as in the seventeenth, there is no substitute for solid fieldwork. But as we have seen, being on-site guarantees only testimony, not accuracy; the latter happens only when as many perspectives as possible are used to decipher place. Never before have we had so many tools at our disposal for depicting places. All of them—from remotely sensed satellite images to on-the-ground impressions—are essential to portraying the region's changing character. But we must be careful in assuming that they are *all* we need, for they only offer limited glimpses of what is there.

In April of 2003, I wandered north from Las Vegas into the heart of the Great Basin. I have taken many similar jaunts for more than forty years, but this time my maps were like those in the TV show *Tremors,* both electronic and paper. In the early 1960s, my best maps would have been USGS topographic maps, and I admit my enduring fondness for these paper artifacts. In fact, I still have many dog-eared veterans of these trips, and they usually accompany me here. By the end of the twentieth century, the USGS's surveying of the Great Basin had resulted in a quadrangular mosaic of highly detailed maps for the entire region. Originally surveyed at a fairly small scale of 1:100,000, then 1:62,500, all of the region is now surveyed at 1:24,000. These USGS "topo sheets" are used not only by resource managers but also by recreationists, including hunters, off-roaders, and solitude seekers like me. The map of Paradise Valley presents a contoured view of an archetypal section of Nevada's basin and range country (fig. 8.15). In the higher country, piñon pines dot the landscape, while trees line the aptly named Big Cottonwood Creek draining toward the Little Humboldt River.

According to the USGS, each topographic map "accurately represents the natural and manmade features of the land." To accomplish this, "the shape and elevation of the terrain are portrayed by contour lines and specific features such as roads, towns, water areas and vegetation are portrayed by map symbols and colors."[16] Reading a topographic map involves both art and science. On one level, the map's contours present an accurate delineation of the earth's surface and reveal the distribution of vegetation. On another, the human imagination translates a density of contour lines into rugged country, and stark white surfaces devoid of contours as nearly smooth land. When read in terms of abstract patterns, topographic maps reveal the Great Basin landscape to be a mosaic of varied features that conspire to create a regional identity. This landscape is understood on numerous levels, including aesthetic, and even this scientific map mirrors this sentiment.

A glimpse at those mostly white maps of the Great Basin reveals how scarce water, vegetation, and human settlements are throughout most of the region. Other USGS maps showing mining districts (for example, the Eureka Mining District, Nevada) and national parks and monuments (for example, Death Valley) are a reminder that special places of economic or aesthetic value draw us into a largely unsettled region. These maps are updated by aerial and satellite information periodically, a reminder that the region is under constant surveillance. The completion of the USGS mapping project was unheralded but noteworthy. The USGS maps are a tribute to the coordination of data from numerous agencies and represent some of the

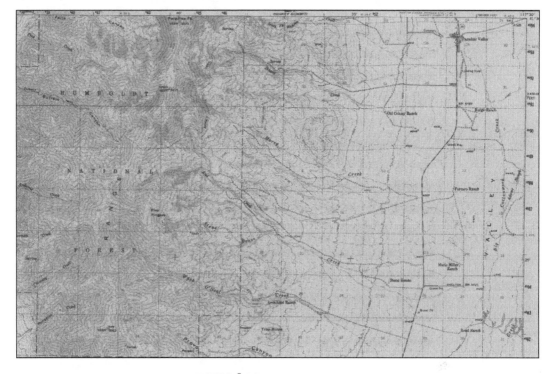

FIGURE 8.15.

Topographic maps by the United States Geological Survey ultimately covered the
entire Great Basin region (Paradise Valley, Nevada, quadrangle). Author's collection

finest extensive mapping ever conducted, printed, and made available to the
public. The Great Basin's coverage alone required about 3,500 separate
maps—a testimony to not only how detailed, but also how fragmented, our
knowledge and maps of the region had become.

The 3,500 USGS topographic maps offer detailed portraits of small portions of the Great Basin. Those wishing to gain a more comprehensive
understanding of topography at the subregional level consult maps done at
about ten times smaller scale—1:250,000—that cover much larger areas.
Originally produced by the Army Map Service, these maps offer a sweeping view of the landscape. In the 1950s and 1960s, advances in injection
molding enabled plastic raised-relief versions of these maps to be produced.
Like the relief map made by the old Indian for J. G. Bruff and his fellow
travelers in 1849, these maps present topography in the third dimension. We
can run our fingers over the map's surface to feel the lay of the land. More
recently, the Jeppeson Map Company (1960s) and Gousha Company (1990s)
produced raised-relief maps of various states. The 1997 raised-relief map of

Nevada by Gousha is a work of art revealing the intricate arrangement of mountains and basins in that state, as is its counterpart for Utah. But alas, these are two separate maps, and the result is that we are denied experiencing the sweep of the Great Basin from the Sierra Nevada to the Wasatch Mountains unless we join the two.

I, like other travelers here, demand more and better information to keep from getting lost when I am out in the field. Detailed maps from the USGS serve this purpose, but they also satisfy my nearly insatiable appetite to know as much as I can about this fascinating place. On this trip, I have not only a batch of USGS topographic maps and a *DeLorme Atlas of Nevada,* but something that now enables me to pinpoint my exact location—DeLorme's Earthmate GPS receiver, which I can use with my laptop computer. The effect is both reassuring and mesmerizing. DeLorme's "Topo USA" software provides detailed 2-D and 3-D maps of the entire nation, including the Great Basin, suggesting that no place is now *terra incognita.* These products that provide ready access to a once-remote region do two things: They help to demystify it by increasing our knowledge of it and also enable us to pinpoint our own position in it. This is what all explorers ultimately sought through the centuries.

And yet I continue to ponder an enduring question that increases in urgency as mapmaking gets better and better: *Is mapping a place in greater and greater detail actually synonymous with really knowing or understanding it?* Here, in the middle of the Great Basin, I conclude that it isn't: Here, with the place itself spread out before me and its features depicted on maps, I realize that enhanced mapping should never be confused with complete knowledge. Until we can map all the other aspects that make up this magnificent place—the erratic motion of a dust devil moving across a playa, the pungent smell of sagebrush wafting on a cool breeze after a rain shower, the sizzling feel of chocolate-colored basalt under a Great Basin summer sun—imagination will always be one step ahead of, and yet will always inspire, mapmaking.

# 9

## Comprehending Cartographic Change

ON FEBRUARY 1, 2003, THE PREDAWN DARKNESS over the southern Great Basin was briefly interrupted by a streak of light that flashed from west to east at 20,000 miles per hour. To those who saw it, the reentry of the space shuttle *Columbia* seemed like a shooting star, but this was normal for a spacecraft entering the outer layers of the atmosphere. Unbeknownst to these observers, however, the *Columbia* was in trouble as it streaked fifty miles high above the California coast and toward the Great Basin. By the time it reached southern Nevada a few seconds later, *Columbia* began to shed its heat-protecting wing tiles—a failure that would lead to the craft's catastrophic disintegration minutes later over Texas. As pieces of the *Columbia* rained down to terra firma, they created a debris field more than a thousand miles long and a hundred miles wide.

Because the first pieces that were shed from the space shuttle could reveal the actual cause of its failure, a massive effort was mounted to locate those parts. However, because they had plummeted to earth in one of the most sparsely populated sections of the country—the southern Great Basin—the search here would be the most difficult. According to NASA representative Kevin O'Toole, the search focused on seven sites—five in Nevada and two in Iron County, Utah—that had the "highest probability" of yielding debris.[1] Thus it was that the first pieces of the shuttle to be lost would be among the last found—a testimony to the enduring isolation and emptiness of the Great Basin to the present day.

Another aspect of that herculean search revealed how much the art and science of cartography has changed in the last century. The exact location of each piece of debris found was carefully identified using GPS. This mapping enabled those reconstructing the final minutes of *Columbia*'s reentry to determine the sequence of destruction by identifying which parts failed first, which failed next, and which failed last, allowing the investigators to reconstruct the process of disintegration. The map of the shuttle's remains took shape within the weeks and months that followed the disaster. This was accomplished much as the map of the Great Basin had taken shape over decades and centuries—a piece here, a revelation there—until the vastness of what was mapped was finally comprehended. These mapped pieces of debris constituted what one observer called the world's largest jigsaw puzzle.

Tellingly, the map of shuttle debris still had many blanks—places of mystery where the mind still sought information that could piece the puzzle together. Like the Great Basin itself in the popular imagination, some places on that debris-field map were destined to remain blank forever—silences, as it were, where pieces could never be encountered, comprehended, and recorded. Those blanks existed for a very good reason: in the fiery disintegration and fall, portions of the space shuttle were vaporized; they had, for all intents, simply vanished without a trace. The missing parts were much like the maps made by earlier peoples—the Indians and nameless travelers whose maps were never recorded and never saved. We suspect, which is to say we intuitively believe, that they existed, but we can never fully understand their role in history. Thus it is that our understanding of the Great Basin's cartographic history is fragmentary. It will be forever incomplete, some of its pages lost to posterity. Like NASA's space shuttle program, mapmaking represents the best effort designers and technicians can summon at the time. Mapmaking and exploration may answer questions about the discovery of places that are very remote, but they always involve human emotions that are surprisingly familiar. In the end, all cartography—like all exploration—attempts to conquer the unfamiliar by making it comprehensible.

Consider, then, the dilemma that has faced all those who sought to map the Great Basin region from 1500 to the present. It was, and remains, a formidable task. The process is ongoing, which is to say the Great Basin's physical and cultural geography presented unique challenges to early explorers and remains a challenge to cartographers to this day because expectations are constantly changing. The region was enigmatic from the beginning of exploration because the Native Americans here did not create permanent, highly visible landmarks like towns, farmlands, and roads (in contradistinction to, say, the pueblos of the desert Southwest). The early explorers had to delineate the region largely on its natural features. Even today, the physical setting dwarfs many of the cultural features, and the region's mountain ranges and hydrology are the benchmarks for most mapmakers. Thus it is that the Great Basin exhibits what Victoria Dickenson calls the "separation between the constructions of men and the works of God." More to the point, this region possesses what Dickenson further characterizes as the "otherness and inherent divinity of the natural world."[2] Yet the landscape here is so stark and immense that it has seemed, to some observers at least, that God himself must have forsaken it. That challenge, which we might say involves "comprehending the incomprehensible," characterized the region's exploration from the outset. Like an entire continent, the Great Basin was both unknown and unknowable until explorers could get a better idea of its outline, then travel into its heart. Significantly, the works of God actually become the works of man, cartographically speaking. People first imagined them, and then placed them at will on maps to fit their preconceptions. Thus it is that natural features with Indian names become familiar Mormon landmarks like Devil's Gate, the Great Stoneface, and Tabernacle Hill.[3]

In our study of maps created over a span of centuries, I believe I have demonstrated that mapping this region was accomplished in fits and starts. Viewed chronologically and comprehensively, in fact, the cartographic history of the Great Basin challenges a simple "progressive" chronological model of increasingly better (that is, more accurate) images. In some cases, we find maps that are clearly retrogressive—that is, less accurate than earlier maps. As historian Jack Jackson has observed, "Maps of the discovery period were merely passed from nation to nation, copied and recopied, until they gained credibility through sheer familiarity."[4] Exactly how and why this process takes place is both fascinating and of intense interest to students of antiquarian maps. On the other hand, some maps, even if in the minority, appear to be far more accurate than their peers. Consider again William Winterbotham's 1795 map, which seems to be years ahead of other British maps in regard to the depiction of the Great Salt Lake. Why does his map break with convention and tradition, offering a surprisingly different and considerably more advanced understanding? In addition to being less en-cumbered by tradition, Winterbotham did not have quite the same agenda as Anglo-Americans. We might say that he had less invested in the idea of a westward-flowing river from the Great Salt Lake to the Pacific than those who occupied the area (Spain) or saw themselves expanding westward across it to remove the Spaniards (Anglo-Americans). Yet even Winterboth-am succumbed to the belief that the West's mountains formed a nexus or knot that separated the waters flowing to the Atlantic and to the Pacific. In this regard, Winterbotham had considerable company. His map and others demonstrate that maps and mapmakers fit into "schools," much as we cate-gorize art history. These schools involve a dominant way of representing reality, and they are never separable from their political or social systems. Yet we must be careful, for if we study even familiar maps closely, we see new things we've overlooked.

Here I should like to call again on the words of Victoria Dickenson, who encountered a similar enigma in studying the natural-history illustration of flora and fauna—which has much in common with cartography. Dickenson isolated three main concerns when interpreting all images that transform unknown nature into the known:

1. *Chronology.* Time is important, and we may recognize either "broad continuities or displacement of one system of thought by another" over time. Although that progression *generally* characterizes the Great Basin over a long period of time, there is a caveat. The fact that images do not always "progress" in an orderly fashion may seem incomprehensible—until, that is, we recall Foucault's statement that "the same, the repetition, and the unin-terrupted are no less problematic than the ruptures."[5]

2. *Purpose.* What, we may ask, is the dominant purpose of a particular car-tographic image? To inform accurately? To claim for posterity? To encour-age trade and commerce? Here we should realize that representations may be "locked into a system of beliefs in which form and meaning might be regarded as inseparable." This in turn means that "accuracy in naturalistic representation might be subordinated to concern for style, [or] for viewer's expectations." Thus it is that a style of representing something "is not sim-ply the external shell under which true meaning is hidden; it is in itself redolent with the artist's [or, I might add, cartographer's] understanding of meaning."[6] Style is, in a word, substance.

3. *Context.* Like any other images, maps can never be separated from the "social and economic conditions of the time and the purposes of the artist or author." Brian Harley expands the idea of contexts to note that maps must be considered in *multiple* contexts, namely the cartographer's context, the context of other maps, and the context of society.[7] More to the point, Harley discusses the immense silences that are embodied in many maps—silences that are so powerful they obscure the voices and beliefs of others.

There is another context that has hardly been touched upon in the study of cartographic history but is central to understanding how maps function. It involves the relationship between the map itself and other imagery. It appears no coincidence that our scientific understanding of the Great Basin's closed or interior hydrology developed at exactly the same time in history that other images—for example, detailed scenes of geological strata or the rendering (and naming) of topographic features like the triangular island in its namesake Pyramid Lake—were made. Those convergences in the 1840s suggest that cartography and natural-history illustration—too long considered separate subjects—are in fact interdependent. It is tempting to ask which influenced which, cartography or illustration, until we realize that they are simply inseparable. The Frémont-Preuss map of the Great Basin is in fact a product of, and in turn influenced, increasing expectations about fidelity to the environment depicted. Of course, all cartography and art—even when done for scientific purposes—is somewhat subjective. This study suggests that the cartographer and illustrator of the 1840s have the same symbiotic relationship that the digital satellite image and website designer have today; they are complementary and interdependent. Viewed thusly, our definitions of a map may need to be expanded to include many tangential but essential subjects, including environmental sculpture, mural art, and the like, for these also provide an orientation to place.[8]

In the twentieth century alone, the geological mapping of the Great Basin reveals how radically perceptions/questions of what is a map—as well as what constitutes the region—have evolved. In the early twentieth century, geological maps of the Great Basin began to reveal the incredible complexity of bedrock geology. The basic outline seems simple enough—a series of mountain ranges lined up with sediment-filled basins between them. But each of those mountains is different; each may consist of rock units as varied as bedded limestones, banded quartzites, and lava flows. These ranges reflect the changes that have gone on over millions of years, but we see them as permanent and enduring. We also see them as two-dimensional until we look more closely. By the late nineteenth century, mining maps enabled us to peer into parts of the earth beneath the Great Basin to get an intimate view. This, however, was not enough. Mining companies next constructed three-dimensional models to describe their workings into various ore bodies. In the early twentieth century, Mrs. Hugh Brown of Tonopah, Nevada, described such a three-dimensional model: "On thin glass slides, some of which hung vertically in slender grooves while others lay horizontally on tiny cleats, all the workings of the mine were traced to scale in colored inks." This we recognize as a map, of course, and it intrigued Mrs. Brown: "When you stood in front of the model and looked into its serried sections, you seemed to be looking into the earth with a magic eye." The model enabled the viewer to see "'drifts' and 'stopes' and 'crosscuts' with every foot of ore blocked out," and on it "one could trace the meandering vein . . . where it petered out or widened into riches unimagined as it continued into regions still unexplored."[9] Unexplored regions no longer lay on the earth's surface as geographical space in the Great Basin at this time because the surface had been mapped. Instead, *terra incognita* now beckoned in the third dimension.

But consider yet another challenge of mapping the geology of the Great Basin: That geology itself is in constant flux and motion. The region is constantly being stretched apart as the Pacific Plate shears. If those fault block mountains on geological maps could be seen over thousands, or hundreds of thousands, of years, they would be trembling and dancing to the irrepressible forces of continental drift. In the vertical dimension, large blocks are being pulled under by subduction; others are rising; in still other cases, magma moves onto the surface by eruption and sheet flow. Horizontally, too, mountains are literally moving, shattering, and being reassembled into new forms.

We can experience this movement by logging onto a website, or viewing a video, developed by geologist Chris Scotese of the University of Texas at Arlington. Using a series of clues, Scotese has constructed a map of the world from the beginning of recorded geological time to the present. On it, we watch the continents fuse and then break apart, only to regroup and then split apart again. The Great Basin emerges over hundreds of millions of years of geological time (about thirty seconds as we watch the map morph on the video screen), until we recognize its familiar shape in the present. But there is more. Based on the parameters that brought the earth's crust to this point, Scotese has projected what will happen over the next fifty million years, and it is even more awe-inspiring. The North American continent's western margin continues its relentless breakup, with much of California shearing off as it slides northwestward. This startling development not only endorses those dark jokes about the Golden State sliding into the sea; it harkens back to cartographers like Coronelli who, in the 1600s and early 1700s, depicted California as an island.

As geological forces continue to reconfigure the continent, the Great Basin also changes dramatically as the geological clock registers about 50 million years into the future. Where the Sierra Nevada once formed the western margin of the Great Basin, tectonic forces ultimately depress the area that once lay several thousand feet above sea level, and the Great Basin sinks to become the floor of a new ocean. On Scotese's map, virtually the entire region has disappeared, becoming undersea topography obscured by a sheet of blue. At that time, if anyone were around to experience it, the area that was once the Great Basin would be the subject of oceanographic mapping; but those would be maps nevertheless. This glimpse of geological mapping confirms not only that maps are reproduced through changing technology, but that the technology itself drives what we map. It is worth restating that cartography is the art and the science of mapmaking. Both work hand in hand to help us depict places in light of changing knowledge about them.

Consider, then, the ultimate paradox of the Great Basin's cartography. To understand it fully requires atomization. We need to decipher it in greater and greater detail—which is to say at a greater cartographic scale—until we come perilously close to losing our comprehension of it as a region. After all, no single thing—not even the region's trademark interior-drained rivers or "Great Basin" sagebrush—is found *only* in the Great Basin, and so we need to keep our eye on the larger picture as we pursue the minutiae. Since coming to understand both the region's perimeter and its interior about 150 years ago, we now see the Great Basin not as a single region at all, but as a series of increasingly complex subregions: hydrographic studies reveal it to be many basins; vegetation studies reveal it to encompass several major plant community regions, in the Great Basin proper and in the Mojave Desert;

and geology links it to forces extending to the rest of the Basin and Range province beyond. The relationship between cartography, technology, and art is enduring, yet always changing as our understanding of the place evolves. With increasingly sophisticated satellite maps and more intensive settlement, one might even predict that the name "Great Basin" itself will become increasingly obsolete, which is to say historical or antiquarian. In a sense, it already has. But we should never underestimate the significance of the old maps on the popular imagination. They play a role in mapmaking and map reading to this day. Their very age and their naive understandings continue to enchant and inform us about people long forgotten and places we still long to understand.

# Epilogue

MAPS HAVE HELPED CREATE, and then sustain, the image of the Great Basin as a place of wonder and mystery for generations. There is evidence that they will continue to do so in the future. In his fascinating book *Geology of the Great Basin,* Bill Fiero hints at some of the region's compelling power as a cartographic icon. "As a child," Fiero begins, "I would trace the outlines with a finger and follow the border of earth's great features on maps." Fiero's inquiring fingertip traversed the maps in search of magical features like the Himalayan Mountains and the icy recesses of the Arctic. "But nowhere," he confesses, "did my finger roam a more mysterious region than the Great Basin." Here his imagination "conjured visions of a vast sandy bowl, replete with rattlesnakes and alkali flats, rimmed totally by lofty mountains." The tactile experience of deciphering the map, and hence the region it represented, was exhilarating: "My finger would quiver as it walked the rimrock around one quarter of a million square miles of the dry bowl."[1]

These are the kinds of images that inspire explorers and authors. Basing his intimate knowledge of the region on hundreds of geological maps prepared over a century or more, Fiero would later draw dozens of sketch maps of a geologically contorted Great Basin for his book. These maps portray a region undergoing a spectacular evolution over eons as mountains rose, seas parted, continents moved. Yet no maps influenced Fiero more than those that first enchanted him as a boy. They were catalytic, galvanizing a young mind and its fertile imagination to wonder, then explore. It is to these maps we first turn, and must always inevitably return, when we conceive of a place and a region. These maps were influential in shaping the popular psyche in the age of exploration. As antiquarian maps today, they convey a sense of naiveté, but astute readers will sense in Fiero's commentary not only the innocence of youth, but also the budding inquisitiveness of adolescence. The roaming finger that Fiero uses so aptly to discuss a youthful enchantment for exploring maps is something much more—a perfect metaphor for both the explorer who points toward the distant horizon and the lover who seeks equally "mysterious" regions of sexual pleasure with a "quivering"—to use Fiero's word—finger. That finger points to the exotic or distant places on the one hand, but is also the stimulator of an internal erotic energy that keeps humankind exploring. That metaphorical finger is a subliminal reminder that exploration seeks both knowledge and intimacy. Like exploration itself, then, mapping is never the innocent process it first seems, for it demands even more knowledge, first of surfaces and then of more hidden places. Nor are the maps produced in the process neutral or innocent, for they work hand in hand with exploration to first intrigue, then inform, and ultimately seduce.

# Notes

INTRODUCTION: MAPS AND MEANING

1. For a penetrating analysis of the relationship between art and cartography, see Edward Casey, *Representing Place: Landscape Painting and Maps* (Minneapolis: University of Minnesota Press, 2002).

2. Bird's-eye views are discussed in David Buisseret, ed., *From Sea Charts to Satellite Images: Interpreting North American History through Maps* (Chicago: University of Chicago Press, 1990).

3. Gary Hausladen, ed., *Western Places, American Myths: How We Think about the West* (Reno: University of Nevada Press, 2003), 7.

4. Donald K. Grayson, *The Desert's Past: A Natural Prehistory of the Great Basin* (Washington, DC: Smithsonian Institution Press, 1993); Stephen Trimble, *The Sagebrush Ocean: A Natural History of the Great Basin* (Reno: University of Nevada Press, 1999, 1989).

5. William L. Fox, *The Void, the Grid, and the Sign: Traversing the Great Basin* (Salt Lake City: University of Utah Press, 2000).

6. Richard Francaviglia, *Believing in Place: A Spiritual Geography of the Great Basin* (Reno: University of Nevada Press, 2003).

7. Phillip Allen, *Mapmaker's Art: Five Centuries of Charting the World: Atlases from the Cadbury Collection, Birmingham Central Library* (New York: Barnes & Noble Books, 2000); John Goss, *The Mapping of North America: Three Centuries of Map-Making, 1500-1860* (Secaucus, NJ: Wellfleet Press, 1990); Edmund William Gilbert, *The Exploration of Western America, 1800-1850* (New York: Cooper Square, 1966).

8. Denis Wood, *The Power of Maps* (New York: Guilford Press, 1992); J. B. Harley, *The New Nature of Maps: Essays in the History of Cartography* (Baltimore: Johns Hopkins University Press, 2001).

9. See Richard Francaviglia, "Exploring Historic Maps On-Line," *Fronteras* 10, no. 1 (spring 2001): 3-4.

## CHAPTER 1. COMPREHENDING THE GREAT BASIN

1. Adobe Workshop and other computer programs greatly facilitate this process.

2. See John G. Mitchell, "The Way West," *National Geographic* 198, no. 3 (September 2000): 55-59.

3. William A. Bell, *New Tracks in North America: A Journal of Travel and Adventure whilst Engaged in the Survey for a Southern Railroad to the Pacific Ocean in 1867-1868* (London: Chapman and Hall, 1870), 487.

4. Amanda M. Brown, "Of Those Who Served: The Veterans History Project Collection at the Library of Congress," *Folklife Center News* 25, no. 2 (winter 2003): 5.

5. Mark Twain, *The Works of Mark Twain,* vol. 2, *Roughing It,* introduction and notes by Frank R. Rogers, text established and textual notes by Paul Baender (Berkeley: Published for the Iowa Center for Textual Studies by the University of California Press, 1972), 148-49.

6. DRI News, http://newsletter.dri.edu/2002/fall/walkerlake.htm.

7. See Paul Carter, "Dark with Excess of Bright: Mapping the Coastlines of Knowledge," in *Mappings,* ed. Denis Cosgrove (London: Reaktion Books, 1999), 125-47.

8. J. H. Simpson, *Report of Explorations . . . 1859* (Reno: University of Nevada Press, 1983), 79.

9. Ibid., 82.

10. Ibid., 87.

## CHAPTER 2. THE POWER OF *TERRA INCOGNITA* (1540-1700)

1. David Weber, *The Spanish Frontier in North America* (New Haven: Yale University Press, 1992), 57.

2. See John Logan Allen, ed., *North American Exploration,* vol. 3, *A Continent Comprehended* (Lincoln: University of Nebraska Press, 1997).

3. Richard Francaviglia, "In Search of Geographical Correctness, or When and Where Maps Lie," *The College* (magazine of the College of Liberal Arts, University of Texas at Arlington), vol. 2, no. 1 (fall 1997), map insert.

4. James Cowan, *A Mapmaker's Dream: The Meditations of Fra Mauro, Cartographer to the Court of Venice* (New York: Warner Books, 1996), 11.

5. Umberto Eco, *The Island of the Day Before* (New York: Harcourt Brace, 1995), 128-29.

6. Cabeza de Vaca's reports, of course, helped motivate Coronado's overland expedition shortly thereafter.

7. See Glyn Williams, *Voyages of Delusion: The Quest for the Northwest Passage* (New Haven: Yale University Press, 2003).

8. W. P. Cumming et al., *The Exploration of North America, 1630-1776* (New York: G. P. Putnam's Sons, 1974),181.

9. See Michael Mathes, "Non-Traditional Armies in New Spain during the Habsburg Viceroyalties and Their Service in Exploratory Expeditions," *Terrae Incognitae* 35 (2003): 16-27, and Richard Flint and Shirley Flint, eds., *The Coronado Expedition to Tierra Nueva: The 1540-1542 Route across the Southwest* (Niwot: University Press of Colorado, 1997).

10. See Dava Sobel, *Longitude: The True Story of a Lone Genius Who Solved the Greatest Scientific Problem of His Time* (New York: Walker, 1995).

11. Joseph Conrad, *Heart of Darkness* (New York: WW Norton, 1988), 11.

12. See Denis Cosgrove, *Apollo's Eye: A Cartographic Genealogy of the Earth in the Western Imagination* (Baltimore: Johns Hopkins University Press, 2001).

13. David Buisseret, *The Mapmaker's Quest: Depicting New Worlds in Renaissance Europe* (New York: Oxford University Press, 2003), 43.

14. See Dennis Reinhartz, *The Cartographer and the Literati: Herman Moll and His Intellectual Circle* (Lewiston, NY: E. Mellen Press, 1997).

15. Dennis Reinhartz, "*Legado:* The Information of the Entradas Portrayed through the Early Nineteenth Century," in *The Mapping of the Entradas into the Greater Southwest,* ed. Dennis Reinhartz and Gerald Saxon (Norman: University of Oklahoma Press, 1998), 132-51.

16. See Paul E. Cohen, "The King's Globemaker," in *Mapping the West: America's Westward Movement, 1524-1890* (New York: Rizzoli International, 2002), 42-47.

17. See Buisseret, *Mapmaker's Quest,* 183-84.

CHAPTER 3. MAPS AND EARLY SPANISH EXPLORATION (1700-1795)

1. Peter Nabokov, "Orientations from Their Side: Dimensions of Native American Cartographic Discourse" in *Cartographic Encounters: Perspectives on Native American Mapmaking and Map Use,* ed. Malcolm Lewis (Chicago: University of Chicago Press, 1998), 243.

2. D. Graham Burnett, *Masters of All They Surveyed: Exploration, Geography, and a British El Dorado* (The University of Chicago Press, Chicago, 2000), 183.

3. Ibid., 186.

4. See Patricia Galloway, "Debriefing Explorers: Amerindian Information in the Delisles' Mapping of the Southeast" in *Cartographic Encounters,* ed. Lewis, 234-

35, and especially G. Malcolm Lewis, "Indicators of Unacknowledged Assimilations from Amerindian Maps on Euro-American Maps of North America: Some General Principals Arising from a Study of LaVérendrye's Composite Map, 1728-29," *Imago Mundi* 38 (1986): 9-34.

5. Lewis, *Cartographic Encounters,* 11.

6. David Woodward and Malcolm Lewis, eds., *The History of Cartography,* vol. 2, book 3, *Cartography in the Traditional African American, Arctic, Australian, and Pacific Societies* (Chicago: University of Chicago Press, 1998), 257.

7. Dennis Reinhartz, "An Exploration of the 'Silences' on Various Late Eighteenth-Century Maps of Northern New Spain," *Terrae Incognitae* 35 (2003): 45.

8. Ibid.

9. This map appears in Jacques Nicolas Bellin, *L'hydrographie françoise* (Paris, 1753-1756).

10. Walter Briggs, *Without Noise of Arms: The 1776 Dominguez-Escalante Search for a Route from Santa Fe to Monterey* (Flagstaff: Northland Press, 1977), 106-8.

11. Briggs, *Without Noise of Arms,* 109-10.

12. Herbert Eugene Bolton, *Pageant in the Wilderness: The Story of the Escalante Expedition to the Interior Basin, 1776, Including the Diary and Itinerary of Father Escalante* (Salt Lake City: Utah State Historical Society, 1950), 245.

13. See also Briggs, *Without Noise of Arms,* 180.

14. See Bolton, *Pageant in the Wilderness.*

15. Carl I. Wheat, *The Spanish Entrada to the Louisiana Purchase, 1540-1804,* vol. 1 of *Mapping the Trans-Mississippi West, 1540-1861* (San Francisco: Institute of Historical Cartography, 1957-1963), 115.

16. This map appears opposite page 70 in Jonathan Carver, *Travels through the Interior Parts of North-America, in the Years 1766, 1767, and 1768* (London: Jonathan Carver, 1778).


CHAPTER 4. IN THE PATH OF WESTWARD EXPANSION (1795-1825)

1. Andrew David et al., eds., *The Malaspina Expedition, 1789-1794: The Journal of the Voyage by Alejandro Malaspina,* vol. 1, *Cadiz to Panama,* with an introduction by Donald Cutter, Hakluyt Society, ser. 3, no. 8 (London: Hakluyt Society, 2002).

2. L. Collin, *Louisiane et pays voisins, d'après les relations et les cartes les plus récentes* [Louisiana and Adjacent Territories, from the Most Recent Accounts of Maps], in Louis Narcisse Baudry de Lozières, *Voyage à la Louisiane, et sur le continent de l'Amérique Septentrionale, fait dans les années 1794 à 1798* (Paris, 1802).

3. Wheat, *Mapping the Trans-Mississippi West,* 134.

4. Alexander von Humboldt, *Political Essay on the Kingdom of New Spain,* vol. 1 (London: Longman, Hurst, Rees, Orme and Brown, 1811), iv-v.

5. Ibid., lxxxiii.

6. Ibid., lxxxiv-lxxxv.

7. Ibid., lxxxvi-lxxxvii.

8. Correspondence between Jefferson and Humboldt, as cited in Helmut De Terra, "Alexander Von Humboldt's Correspondence with Jefferson, Madison, and Gallatin," *Proceedings of the American Philosophical Society* 105, no. 6 (December 1959): 792-94.

9. Ibid.

10. Spain had, in fact, hoped to intercept Lewis and Clark by force but was not able to—thereby conceding, as it were, to the ambitious Jefferson.

11. Paul E. Cohen, *Mapping the West: America's Westward Movement, 1524-1890* (New York: Rizzoli International, 2002), 80.

12. John Logan Allen, "Thomas Jefferson and the Mountain of Salt: A Presidential Image of Louisiana Territory," *Historical Geography,* Special Louisiana Purchase Bicentennial Edition, 31 (2003): 13.

13. The map appears opposite page 1 in Hubbard Lester, comp., *The Travels of Capts. Lewis & Clark from St. Louis, by Way of the Missouri and Columbia River, to the Pacific Ocean* (London: Longman, Hurst, Reese, and Orme, 1809).

14. The map, *Carte pour servir au Voyage des Capes. Lewis et Clarke à l'Océan Pacifique,* appeared in Patrick Gass, *Voyage des Capitaines Lewis et Clarke depuis l'embouchure du Missouri, jusqú à l'entrée de la Columbia dan l'Océan Pacifique....* (Paris: Arthus Bertrand, 1810).

15. Melish's map was meant to accompany his book *A Geographical Description of the United States, with the Contiguous British and Spanish Possessions, with the Contiguous British and Spanish Possessions, Intended as an Accompaniment to Melish's Map of These Countries* (Philadelphia, 1816).

16. Ibid., 31.

17. Ibid., 32.

18. Map in the David Rumsey Collection, as illustrated in Cohen, *Mapping the West,* 102-3.

19. See Robert Mills, *Inland Navigation: Plan for a Great Canal between Charleston and Columbia, and for Connecting Our Waters with Those of the Western Country* (Columbia, SC: Telescope Press, 1821), 55-58, which contains a foldout of J. Melish's map, published 1816, improved to 1820 communication with the Pacific with steamboat and rail lines.

20. Seymour I. Schwartz, *The Mismapping of America* (Rochester, NY: University of Rochester Press, 2003).

21. Burnett, *Masters of All They Surveyed,* 84.

22. Françoise Weil, "Relation de voyage: document anthropologique ou text lit-téraire," in *Histoire de l'anthropologie: XVI–XIXe siècles,* ed. Britta Rupp-Eisenreich (Paris: Klincksieck, 1984), 57. See also Christophe Boucher, "'The Land God Gave to Cain': Jacques Cartier Counters the Mythological Wild Man in Labrador," *Terrae Incognitae: The Journal of the Society for the History of Discoveries* vol. 35 (2003): 28-42.

23. Allen, "Thomas Jefferson and the Mountain of Salt," 9.

24. See Elizabeth John, "The Riddle of Mapmaker Juan Pedro Walker," in Stanley Palmer and Dennis Reinhartz, eds. *Essays on the History of North American Discovery* (College Station: Published by the Texas A&M University Press for the University of Texas at Arlington, 1988), 102-28.

25. John, "Riddle of Mapmaker Juan Pedro Walker," 127.

26. Wheat, *Mapping the Trans-Mississippi West,* vol. 2, 64; see also John, "Riddle of Mapmaker Juan Pedro Walker," 127.

## CHAPTER 5. DEMYSTIFYING *TERRA INCOGNITA* (1825-1850)

1. *Niles' Register,* December 1826, as reported in Gloria Griffen Cline, *Exploring the Great Basin* (Norman: University of Oklahoma Press, 1963), 6.

2. Richard Francaviglia and Jimmy L. Bryan Jr., "'Are We Chimerical in This Opinion?' Visions of a Pacific Railroad and Western Expansion before 1845," *Pacific Historical Review* 71, no. 2 (May 2002): 179-202.

3. William Goetzmann, "Seeing and Believing: The Explorer and the Visualization of Place," in *Essays on the History of North American Discovery and Exploration,* ed. Stanley Palmer and Dennis Reinhartz (College Station: Texas A&M University Press for the University of Texas at Arlington, 1988), 135.

4. See Stewart Winger, *Lincoln, Religion, and Romantic Cultural Politics* (DeKalb: Northern Illinois University Press, 2003), 18-19.

5. Washington Irving, *The Adventures of Captain Bonneville, U.S.A., in the Rocky Mountains and the Far West,* ed. Edgeley W. Todd (Norman: University of Oklahoma Press, 1961), 285.

6. Ibid., 160.

7. Nevada, it should be noted, is the nation's most mountainous state.

8. Dale L. Morgan, *Jedediah Smith and the Opening of the West* (Lincoln: University of Nebraska Press, 1965), 11; see also 371-72.

9. Nebenzahl quoted in Cohen, *Mapping the West,* 117-19.

10. See also "Myths and Maps: Making Explorers and Empires," chap. 2 in Burnett, *Masters of All They Surveyed,* 25-66.

11. See Timothy J. Knab, ed. and comp., and Thelma D. Sullivan, trans., *A

*Scattering of Jades: Stories, Poems, and Prayers of the Aztecs* (Tucson: University of Arizona Press, 2003) 21-22.

12. John Charles Frémont, *Geographical Memoir upon Upper California* (Washington, DC: Tippin & Streeper, 1849), 7.

13. John Charles Frémont, *Report of the Exploring Expedition to the Rocky Mountains in the Year 1842, and to Oregon and North California in the Years 1843-44* (Washington, DC: Blair and Rives, 1845), 206.

14. Ibid., 219; also cited in Robert Heizer, "Aboriginal California and Great Basin Cartography," *Reports of the University of California Archaeological Survey* no. 41, January 2, 1958, 2.

15. Ron Tyler, "Prints vs Photographs, 1840-1860" in *Perpetual Mirage: Photographic Narratives of the Desert West,* ed. May Castleberry (New York: Whitney Museum of Art, 1996), 41-48.

16. Ibid., 48.

17. Clifford L. Stott, *Search for Sanctuary: Brigham Young and the White Mountain Expedition* (Salt Lake City: University of Utah Press, 1984), 110.

18. See John M. Fletcher et al., "Cretaceous Arc Tectonism in the Mojave Block: Profound Crustal Modification That Controlled Subsequent Tectonic Regions" in *Geologic Evolution of the Mojave Desert and Southwestern Basin and Range,* ed. Allen Glazner, Douglas Walker, and John M. Bartley, Memoir 195 (Boulder, CO: Geological Society of America, 2002), 131-51.

19. William W. Slaughter and Michael Landon, *Trail of Hope: The Story of the Mormon Trail* (Salt Lake City: Shadow Mountain, 1997), 50.

20. Society for the Diffusion of Useful Knowledge, *Central America II, Including Texas, California, and the Adjacent States of Mexico* (London, 1846).

21. Cohen, *Mapping the West,* 135-38.

22. Published by J. Grassel for Joseph Meyer. This map appeared in the *Neuster Zeitungs-Atlas* (Hildburghausen, Germany: 1852).


CHAPTER 6. MAPS IN THE SAND (1850-1865)


1. William Lewis Manly, *Death Valley in '49: Important Chapter of California Pioneer History* (San Jose, CA: Pacific Tree and Vine, 1894; reprinted, Ann Arbor: University Microfilms, 1966), 98-99.

2. Ibid., 109-10.

3. Ibid., 154, 164.

4. Ibid., 154.

5. This map is item #HM50895, Theodore Palmer Collection, Huntington Library, San Marino, CA.

6. J. G. Bruff, *Gold Rush: The Journals, Drawings, and Other Papers of J. Goldsborough Bruff . . . April 2, 1849-July 20, 1851,* ed. Georgia Willis Read and Ruth Gaines (New York: Columbia University Press, 1949), 453-54.

7. These maps are in the Special Collections Division, University of Texas at Arlington Libraries.

8. William B. Smart and Donna T. Smart, *Over the Rim: The Parley P. Pratt Exploring Expedition to Southern Utah, 1849-50* (Logan: Utah State University Press, 1999), 41-42.

9. Cohen, *Mapping the West,* 164-66.

10. U.S. War Department, *Reports of Explorations and Surveys, to Ascertain the Most Practicable and Economical Route for a Railroad from the Mississippi River to the Pacific Ocean. Made under the Direction of the Secretary of War, in 1853-54,* vol. 1 (Washington, DC: A. O. P. Nicholson, 1855), 56-61.

11. U.S. War Department, *Explorations for a Railroad Route from the Mississippi River to the Pacific, 1853-1855,* vol. 12, book 1.

12. *Map of the Territory of the United States from the Mississippi to the Pacific Ocean: Ordered by the Hon. Jeff'n Davis, Secretary of War to Accompany the Reports of the Explorations for a Pacific Railroad. Scale 1:3,000,000; Northern Section,* in U.S. Senate, 36th Cong., 2d sess., *Reports of Explorations and Surveys to Ascertain the Most Practicable and Economical Route for a Railroad* (Washington, DC: George Bowman, Printer, 1861).

13. Amy DeRogatis, *Moral Geography: Maps, Missionaries, and the American Frontier* (New York: Columbia University Press, 2003), 44.

14. Map No. 2, *From Great Salt Lake to the Humboldt Mountains. From Explorations and Surveys Made under the Direction of the Hon. Jefferson Davis Secretary of War,* by Capt. E. G. Beckwith 3rd Artillery E.[*sic*] W. Egloffstein Topographer for the Route, 1855. Scale of 12 miles to one inch or 1:760,320.

15. Simpson, *Report of Explorations . . . 1859,* 75.

16. Charles Merriam, "Paleozoic Rocks of Antelope Valley, Eureka and Nye Counties, Nevada," Geological Survey Professional Paper 423 (Washington: U.S. Government Printing Office, 1963), 5.

17. Twain, *Roughing It,* 159.

18. Paul F. Starrs, "Connecting the Continent: Esmeralda County, Nevada, and the Atlantic and Pacific Railroad Survey," *Nevada Historical Quarterly* 40, no. 3 (fall 1997): 237-38.

19. Richard Lingenfelter, *Death Valley and the Amargosa: A Land of Illusion* (Berkeley: University of California Press, 1986), 82-83.

20. Richard F. Burton, *The City of the Saints and across the Rocky Mountains to California* (1861; reprinted, New York: Alfred A. Knopf, 1963), 369.

21. Ibid., 1861.

22. George M. Wheeler, *Report upon United States Geographical Surveys West of the One Hundredth Meridian* (Washington, DC: Government Printing Office, 1889), 612.

23. Ibid., 613.

24. Simon Ryan, "Inscribing the Emptiness: Cartography, Exploration, and the Construction of Australia" in *De-Scribing Empire: Post Colonialism and Textuality,* ed. Chris Tiffin and Alan Lawson (London: Routledge, 1994), 116.

25. See Michael Landon, ed., *The Journals of George A. Cannon—Vol. 1—To California in '49* (Salt Lake City: Deseret Book Company, 1999).

26. See Stott, *Search for Sanctuary,* 96-97, 135.

27. Ibid., 109.

28. Cohen, *Mapping the West,* 180.

29. Jock Taylor, *One Hundred Years Ago in Nevada* (Reno: Western Sales Distributing, 1964), 180.

CHAPTER 7. FILLING IN THE BLANKS (1865-1900)

1. Wheeler, *Report upon United States Geographical Surveys,* vol. 1, Appendix F, Geographical Report, 653.

2. Ibid., 654.

3. William A. Bell, *New Tracks in North America: A Journal of Travel and Adventure whilst Engaged in the Survey for a Southern Railroad to the Pacific Ocean in 1867-1868,* with a foreword by Robert O. Anderson (Albuquerque: Horn and Wallace, 1965), lxiv.

4. Ibid., lxiv-lxv.

5. Clarence King in the U.S. Geological Survey 1st Annual Report, 1880, 4; see also "The United States Geological Survey, 1879-1989," http://pubs.usgs.gov/circ/c1050/indexs.htm, consulted November 18, 2003.

6. See Peter Marzio, *Chromolithography, 1840-1900: The Democratic Art* (Boston: David R. Goldine, 1979).

7. Clarence King, "The Comstock Lode," in *Mining Industry,* by James D. Hague with geological contributions by Clarence King (Washington: Government Printing Office, 1870), 18.

8. Clarence King, 1876 Atlas: U.S. Geological Exploration [Survey], 40th parallel, map 4.

9. William Lee Stokes, *Geology of Utah* (Salt Lake City: Utah Museum of Natural History and Utah Geological and Mineral Survey, 1986), 19.

10. Merriam, "Paleozoic Rocks of Antelope Valley," 2.

11. *Map Showing Detailed Topography of the Country Traversed by the Reconnaissance Expedition through Southern & Southeastern Nevada in Charge of Lieut. Geo. M. Wheeler, U.S. Engineers. Assisted by O. W. Lockwood, Corps of Engineers U.S.A. P. W. Hamel Chief Topographer and Draughtsman.* (New York: New York Lithography, Engraving & Printing, 1869). Wheeler is normally credited as the author of this map, and his effort was indeed critical as it is based on his surveys and field mapping. However, the fact that P. W. Hamel is formally identified as the expedition's chief topographer and draughtsman serves as yet another reminder that mapping is rarely an individual effort. A project like Wheeler's elaborate mapping of southern and southeastern Nevada involves considerable teamwork. Without Wheeler's vision and tenacity, the expedition would not have been possible, nor the area mapped, but without Hamel's cartographic imagination and technical skill, the resulting map would have been quite different—likely less visually arresting and exciting and, therefore, less memorable to future generations.

12. Doris Ostrander Dawdy, *George Montague Wheeler: The Man and the Myth* (Athens: Ohio University Press, 1993), 1-2.

13. G. M. Wheeler, "Facts Regarding the Government Land and Marine Surveys of the United States," extract from *Report on the Third International Geographical Congress* (Washington, DC: Government Printing Office, 1885), 467.

14. Dawdy, *George Montague Wheeler,* 74.

15. Ibid., 79.

16. See Richard Lingenfelter, *Death Valley and the Amargosa: A Land of Illusion* (Berkeley: University of California Press, 1986), 94-96. Lingenfelter quotes Captain Egbert from the *Inyo Independent,* Independence, CA, November 25, 1871, and January 27, 1872.

17. For a more detailed interpretation of the differences between mapping and surveying, see Burnett, *Masters of All They Surveyed,* 21-86.

18. Bell, *New Tracks in North America* (1870), 485; for additional information on the railroads of the Great Basin, see David Myrick, *Railroads of Nevada and Eastern California,* 2 vols. (Reno: University of Nevada Press, 1992).

19. Phil Robinson, *Sinners and Saints.*

20. Samuel Bowles, *Our New West: Records of Travel between the Mississippi River and the Pacific Ocean. Over the Plains—Over the Mountains—Through the Great Interior Basin . . .* (Hartford, CT: Hartford Publishing, 1869), 295-96.

21. Harlan D. Unrau, *Basin and Range: A History of Great Basin National Park* (Washington, DC: U.S. Department of the Interior, 1990), 89.

22. See, e.g., James Underhill, *Mineral Land Surveying* (Denver: Mining Reporter Publishing, 1906); Frederick Winiberg, *Metalliferous Mine Surveying*

(London: Mining Publications, 1950); and M. H. Haddock, *The Location of Mineral Fields* (Stationers' Hall Court: Crosby Lockwood and Son, 1926).

23. For more information on pocket maps, see Brian McFarland, "From Publisher to Pocket: Interpreting Early Nineteenth Century American History through the Pocket Maps of Samuel Augustus Mitchell" (master's thesis, University of Texas at Arlington, 2002).

24. Engineer Department, U.S. Army, *Report upon United States Geographical Sketches West of the One Hundredth Meridian in charge of First Lieut. Geo. M. Wheeler,* vol. 2, *Astronomy and Barometric Hypsometry* (Washington, DC: Government Printing Office, 1877).

25. Richard Francaviglia, "Maps and Mining: Some Historical Examples from the Great Basin," *Mining History Journal* 8 (2001): 66-82.

26. See Fox, *The Void, the Grid, and the Sign.*

CHAPTER 8. MAPS OF THE MODERN / POSTMODERN GREAT BASIN
(1900-2005)

1. Curt McConnell, *Coast to Coast by Automobile: The Pioneering Trips, 1899-1908* (Stanford, CA: Stanford University Press, 2000), 153.

2. Ibid., 157.

3. Ibid., 164.

4. Virginia Rishel, *Wheels to Adventure: Bill Rishel's Western Routes* (Salt Lake City: Howe Brothers, 1983), see esp. 53-64.

5. Robert French, "On Board Navigation Systems for Automobiles," paper presented at the Mapping the Earth and Seas Conference, University of Texas at Arlington, November 7, 1997.

6. James Akerman, "Riders Wanted: Maps As Promotional Tools in the American Transportation Industry," paper presented at the Second Biennial Virginia Garrett Lectures on Maps and Popular Culture, University of Texas at Arlington, October 6, 2000.

7. See Douglas A. Yorke Jr. and John Margolies, *Hitting the Road: The Art of the American Road Map* (San Francisco: Chronicle Books, 1996).

8. Kit Goodwin, "All Aboard! Railroad Maps in the Late Nineteenth Century," *Compass Rose* 5, no. 2 (fall 1991).

9. *Map and Log, TWA—The Transcontinental Airline, Shortest—Fastest—Coast to Coast, Route of the Sky Chief* (n.p.: Transcontinental & Western Air, 1939), University of Texas at Dallas, History of Aviation Collection.

10. United Airlines, *Maps of the Main Line Airway* (n.p.: n.p., ca. 1942), University of Texas at Dallas, History of Aviation Collection.

11. Ibid.

12. *American Airlines—Route of the Astrojets—System Map* (Chicago: Rand McNally, n.d. [ca. 1960]) University of Texas at Dallas, History of Aviation Collection.

13. Website, http://www.area51zone.com/base.shtml, accessed January 24, 2003.

14. *Tremors: The Series,* SciFi Channel, premier episode, March 28, 2003.

15. Ibid., second episode, March 28, 2003.

16. USGS, *Nevada. Index to Topographic and Other Map Coverage* (n.p.: USGS, n.d.).

## CHAPTER 9. COMPREHENDING CARTOGRAPHIC CHANGE

1. Jennifer Weaver, "Search for Shuttle Debris Resumes: Agencies Scour Mountains Near Nevada Border," *Cedar City Daily News,* Sunday, April 13, 2003, A1, A10.

2. Victoria Dickenson, *Drawn from Life: Science and Art in the Portrayal of the New World* (Toronto: University of Toronto Press, 1998), 230.

3. See Francaviglia, *Believing in Place.*

4. Jack Jackson, *Flags along the Coast: Charting the Gulf of Mexico, 1519-1759: A Reappraisal* (Austin: Book Club of Texas, 1995), 11.

5. Dickenson, *Drawn from Life,* 237-38.

6. Ibid., 240.

7. See J. B. Harley, "Text and Contexts in the Interpretation of Early Maps" in *From Sea Charts to Satellite Images: Interpreting North American History through Maps,* ed. David Buisseret (Chicago: University of Chicago Press, 1990), 3-15.

8. See Denis Cosgrove, ed., *Mappings* (London: Reaktion Books, 1999).

9. Mrs. Hugh Brown, *Lady in Boomtown: Miners and Manners on the Nevada Frontier* (Palo Alto, CA: American West, 1968).

## EPILOGUE

1. Bill Fiero, *Geology of the Great Basin* (Reno: University of Nevada Press, 1986), 6.

# Bibliography

Allen, John Logan, ed. *North American Exploration*. 3 Vols. Lincoln: University of Nebraska Press, 1997.

Allen, Phillip. *Mapmaker's Art: Five Centuries of Charting the World: Atlases from the Cadbury Collection, Birmingham Central Library*. New York: Barnes & Noble Books, 2000.

Ansari, Mary B. *Nevada Collections of Maps and Aerial Photographs*. Camp Nevada Monograph No. 2. Reno: Camp Nevada, 1975.

*Auto Trails and Commercial Survey of the United States*. Chicago: George F. Cram Company, 1922.

Bargrow, Leo. *History of Cartography*, ed. R. A. Skelton. Rev. ed. Cambridge: Harvard University Press, 1964.

Bartlett, Richard A. *Great Surveys of the American West*. Norman: University of Oklahoma Press, 1962.

Bricker, Charles. *Landmarks of Mapmaking*. Brussels: International Book Society, Time-Life Books, 1968.

Brooks, George R., ed. *The Southwest Expedition of Jedediah S. Smith: His Personal Account of the Journey to California 1826-1827*. Lincoln: University of Nebraska Press, 1989.

Buisseret, David, ed. *From Sea Charts to Satellite Images: Interpreting North American History through Maps*. Chicago: University of Chicago Press, 1992.

Burden, Philip D. *The Mapping of Northern America: A List of Printed Maps, 1511-1670*. Richmansworth, Herts., U.K.: Raleigh, 1996.

Burton, Richard F. *The City of the Saints and across the Rocky Mountains to California*. 1861. Reprinted, New York: Alfred A. Knopf, 1963.

Campbell, John. *Introductory Cartography.* Englewood Cliffs, NJ: Prentice-Hall, 1984.

Carmack, Noel A. "Running the Line: James Henry Martineau's Surveys in Northern Utah, 1860-1882." *Utah Historical Quarterly* 68, no. 4 (fall 2000): 313-31.

Casey, Edward. *Representing Place: Landscape Painting and Maps.* Minneapolis: University of Minnesota Press, 2002.

Charlet, David Alan. *Atlas of Nevada Conifers: A Phytogeographic Reference.* Reno: University of Nevada Press, 1996.

Cline, Gloria Griffen. *Exploring the Great Basin.* Norman: University of Oklahoma Press, 1963.

Cowan, James. *A Mapmaker's Dream: The Meditations of Fra Mauro, Cartographer to the Court of Venice.* New York: Warner Books, 1996.

Crane, Nicholas. *Mercator: The Man Who Mapped the Planet.* New York: H. Holt, 2003.

Cumming, William P., S. E. Hillier, D. B. Quinn, and Glyndwr Williams. *The Exploration of North America, 1630-1776.* New York: G. P. Putnam's Sons, 1974.

Danzer, Gerald A. *Discovering American History through Maps and Views.* New York: Harper Collins, 1991.

DeLyser, Dydia. "'A Walk through Old Bodie': Presenting a Ghost Town in a Tourism Map." In *Mapping Tourism,* ed. Stephen P. Hanna and Vincent J. Del Casino Jr. Minneapolis: University of Minnesota Press, 2003, 79-107.

Eco, Umberto. *The Island of the Day Before.* New York: Harcourt Brace, 1995.

Ehrenberg, Ralph. "Taking the Measure of the Land." *Prologue* 9, no. 3 (fall 1977): 129-50.

Ferris, Warren Angus. *Life in the Rocky Mountains: A Diary of Wanderings on the Sources of the Rivers Missouri, Columbia, and Colorado 1830–1835.* LeRoy Haffen, ed. Denver: The Old West Publishing Company, 1983.

Flint, Richard, and Shirley Flint, eds. *The Coronado Expedition to Tierra Nueva: The 1540-1542 Route across the Southwest.* Niwot: University Press of Colorado, 1997.

Fox William L. *The Void, the Grid and the Sign: Traversing the Great Basin.* Salt Lake City: University of Utah Press, 2000.

Francaviglia, Richard. *Believing in Place: A Spiritual Geography of the Great Basin.* Reno: University of Nevada Press, 2003.

———. "Exploring Historic Maps On-Line." *Fronteras* 10, no. 1 (spring 2001): 3-4.

———. "In Search of Geographical Correctness, or When and Where Maps Lie." *The College* (magazine of the College of Liberal Arts, University of Texas at Arlington), vol. 2, no. 1 (fall 1997), map insert.

———. "Landmarks in the Cartographic History of the Great Basin." *Fronteras* 13, no. 1 (spring 2004): 3-4.

———. "Maps and Mining: Some Historical Examples from the Great Basin." *Mining History Journal* 8 (2001): 66-82.

Francaviglia, Richard, and Jimmy L. Bryan Jr. "'Are We Chimerical in This Opinion?' Visions of a Pacific Railroad and Western Expansion before 1845." *Pacific Historical Review* 71, no. 2 (May 2002): 179-202.

Frémont, John Charles, *Narrative of the Exploring Expedition to the Rocky Mountains in the Year 1842, and to Oregon and North California in the Years 1843-44.* New York: A. S. Barnes, 1847.

———. *Report of the Exploring Expedition to the Rocky Mountains in the Year 1842, and to Oregon and North California in the Years 1843-44.* Washington, DC: Blair and Rives, 1845.

Friis, Herman R. "Highlights in the First 100 Years of Surveying and Mapping and Geographical Exploration of the United States by the Federal Government, 1775-1880." *Surveying and Mapping* 18 (1958): 185-206.

Gilbert, Edmund William. *The Exploration of Western America, 1800-1850.* New York: Cooper Square, 1966.

Goetzmann, William H. *Army Exploration in the American West, 1803-1863.* Lincoln: University of Nebraska Press, 1979.

———. *Exploration and Empire: The Explorer and the Scientist in the Winning of the American West.* Austin: Texas State Historical Association, 1993.

Goss, John. *The Mapping of North America: Three Centuries of Map-Making, 1500-1860.* Secaucus, NJ: Wellfleet Press, 1990.

———. *The Mapmaker's Art: An Illustrated History of Cartography.* London: Studio Editions, 1993.

Grayson, Donald K. *The Desert's Past: A Natural Prehistory of the Great Basin.* Washington, DC: Smithsonian Institution Press, 1993.

Harley, J. B. (John Brian). *The New Nature of Maps: Essays in the History of Cartography.* Baltimore: Johns Hopkins University Press, 2001.

Harrisse, Henry. *The Discovery of North America: A Critical Documentary and Historical Investment with an Essay on Early Cartography of the New World. . . .* Reprint. Amsterdam: N. Israel, 1969.

Hausladen, Gary, ed., *Western Places, American Myths: How We Think about the West.* Reno: University of Nevada Press, 2003.

Irving, Washington. *The Adventures of Captain Bonneville, U.S.A., in the Rocky Mountains and the Far West.* Ed. Edgeley W. Todd. Norman: University of Oklahoma Press, 1961.

John, Elizabeth Ann Harper. *Storms Brewed in Other Men's Worlds: The Confrontation of Indians, Spanish, and French in the Southwest, 1540-1795.* 2d ed. Norman: University of Oklahoma Press, 1996.

Johnson, Adrian. *America Explored: A Cartographical History of the Exploration of North America.* New York: Viking Press, 1974.

Karrow, Robert. *Mapmakers of the Sixteenth Century and Their Maps.* Chicago: Published for The Newberry Library by Speculum Orbis Press, 1993.

Kish, George. *The Discovery and Settlement of North America 1500-1865: A Cartographic Perspective.* New York: Harper and Row, 1978.

Knoepp, D. P., ed. *Exploration and Mapping of the American West: Selected Essays.* Occasional Paper No. 1, Map and Geography Round Table of the American Library Association. Chicago: Speculum Orbis Press, 1986.

Manasek, F. J. *Collecting Old Maps.* Norwich, VT: Terra Nova Press, 1998.

Martin, James C., and Robert Sidney Martin. *Maps of Texas and the Southwest, 1513-1900.* Austin: Texas State Historical Association, 1998.

Martin, Robert Sidney, and James C. Martin. *Contours of Discovery: Printed Maps Delineating the Texas and Southwestern Chapters in the Cartographic History of North America.* Austin: Texas State Historical Association, 1981-82.

Mathes, Michael. "Non-Traditional Armies in New Spain during the Habsburg Viceroyalties and Their Service in Exploratory Expeditions." *Terrae Incognitae* 35 (2003): 16-27.

Mitchell, John G. "The Way West." *National Geographic* 198, no. 3 (September 2000): 58-59.

Modelski, Andrew M. *Railroad Maps of North America: The First Hundred Years.* Washington, DC: Library of Congress, 1984.

Moffat, Riley Moore. *Printed Maps of Utah to 1900: An Annotated Cartobibliography.* Santa Cruz, CA: Western Association of Map Libraries, 1981.

Monmonier, Mark S. *How to Lie with Maps.* Chicago: University of Chicago Press, 1996.

——. *Map Appreciation.* Englewood Cliffs, NJ: Prentice-Hall, 1988.

——. *Maps, Distortion, and Meaning.* Washington, DC: Association of American Geographers, 1977.

——. *Technological Transition in Cartography.* Madison: University of Wisconsin Press, 1985.

Morgan, Dale L. "Early Maps." *Nevada Highways and Parks* (Centennial Issue 1964): 11-17.

——. *The Humboldt, Highroad of the West.* New York: Farrar & Rinehart, 1943.

——. *Jedediah Smith and the Opening of the West.* Lincoln: University of Nebraska Press, 1965.

——. *The West of William H. Ashley: The International Struggle for the Fur Trade of the Missouri, the Rocky Mountains, and the Columbia, with Explorations Beyond the*

*Continental Divide, Recorded in the Diaries and Letters of William H. Ashley and His Contemporaries, 1822-1838.* Denver: Old West, 1964.

Myrick, David F. *Railroads of Nevada and Eastern California.* 2 vols. Berkeley, CA: Howell-North Books, 1962-63.

Neihardt, John G. *The Splendid Wayfaring: Jedediah Smith and the Ashley-Henry Men, 1822-1831.* Lincoln: University of Nebraska Press, 1970.

Norcross, C. A. *Nevada Agriculture.* San Francisco: Southern Pacific Homeseeker's Bureau, 1911.

Parker, Edmond T., and Michael P. Conzen. "Using Maps as Evidence: Lessons in American Social and Economic History." ERIC Document Reproduction Service, ED 125 935, 1975.

Peterson, Michael P., ed. *Maps and the Internet.* London: Elsevier Science, 2003.

Phillips, Fred M. *Desert People and Mountain Men: Exploration of the Great Basin, 1824-1865.* Bishop, CA: Chalfant Press, 1977.

Reinhartz, Dennis, and Charles Colley. *The Mapping of the American Southwest.* College Station: Texas A&M University Press, 1987.

Reinhartz. Dennis, and Gerald Saxon. *The Mapping of the Entradas into the Greater Southwest.* Norman: University of Oklahoma Press, 1998.

Rishel, Virginia. *Wheels to Adventure: Bill Rishel's Western Routes.* Salt Lake City: Howe Brothers, 1983.

Ristow, Walter William. *American Maps and Mapmakers: Commercial Cartography in the Nineteenth Century.* Detroit: Wayne State University Press, 1985.

Schwartz, Seymour I. *The Mismapping of America.* Rochester, NY: University of Rochester Press, 2003.

Schwartz, Seymour I., and Ralph E. Ehrenberg. *The Mapping of America.* New York: Harry N. Abrams, 1980.

Simpson, J. H. [James Hervey]. *Report of Explorations across the Great Basin of the Territory of Utah for a Direct Wagon-Route from Camp Floyd to Genoa, in Carson Valley, in 1859.* 1876. Reprinted, Reno: University of Nevada Press, 1983.

Skelton, R. A. *Explorers' Maps: Chapters in the Cartographic Record of Geographical Discovery.* New York: Frederick A. Praeger, 1958.

Skelton, R. A. *Maps: A Historical Survey of Their Study and Collecting.* Chicago: University of Chicago Press, 1975.

Sobel, Dava. *Longitude: The True Story of a Lone Genius Who Solved the Greatest Scientific Problem of His Time.* New York: Walker, 1995.

Southworth, Michael, and Susan Southworth. *Maps: A Visual Survey and Design Guide.* Boston: Little, Brown, 1982.

Starrs, Paul F. "Connecting the Continent: Esmeralda County, Nevada, and the

Atlantic and Pacific Railroad Survey." *Nevada Historical Society Quarterly* 40, no. 3 (fall 1997): 232-52.

—. "Esmeralda County, Nevada: Empty Land? Poor Land? Fair Land?" Master's thesis, University of California-Berkeley, 1984.

—. "Home Ranch: Ranchers, the Federal Government, and the Partitioning of Western North American Rangeland." Ph.D. diss., University of California-Berkeley, 1989.

Thrower, Norman J. W. *Maps and Civilization: Cartography in Culture and Society.* Chicago: University of Chicago Press, 1996.

—. *Maps and Man: An Examination of Cartography in Relation to Culture and Civilization.* Englewood Cliffs, NJ: Prentice-Hall, 1972.

Trimble, Stephen. *The Sagebrush Ocean: A Natural History of the Great Basin.* Reno: University of Nevada Press, 1989.

Twain, Mark. *The Works of Mark Twain.* Vol. 2. *Roughing It.* Introduction and notes by Frank R. Rogers, text established and textual notes by Paul Baender. Berkeley: Published for the Iowa Center for Textual Studies by the University of California Press, 1972.

U.S. War Department. *Explorations and Surveys for Rail Road Routes from the Mississippi River to the Pacific Ocean. War Department. Profiles of the Main Routes Surveyed, compiled in 1855.* Washington, DC: U.S. War Department, 1856.

U.S. War Department. *Reports of Explorations and Surveys, to Ascertain the Most Practicable and Economical Route for a Railroad from the Mississippi River to the Pacific Ocean. Made under the Direction of the Secretary of War, in 1853-54.* Vol. 1. Washington, DC: A. O. P. Nicholson, Printer [etc.], 1855-60.

U.S. War Department. *Reports of Explorations and Surveys, to Ascertain the Most Practicable and Economical Route for a Railroad from the Mississippi River to the Pacific Ocean.* Washington, DC: Beverly Tucker, etc., 1855-61.

U.S. War Department. *Reports of the Secretary of War, communicating copies of all reports of the engineers and other persons, employed to make explorations and surveys to ascertain the most practicable and economical route for a railroad from the Mississippi River to the Pacific Ocean, that have been received at the department.* Washington, DC: n.p., 1854.

Van Noy, Rick. *Surveying the Interior: Literary Cartographers and the Sense of Place.* Reno: University of Nevada Press, 2003.

Weaver, Glen D. "Nevada's Federal Lands." *Annals of the Association of American Geographers* 59, no. 1 (March 1969): 27-49.

Weber, David. *The Spanish Frontier in North America.* New Haven: Yale University Press, 1992.

Wheat, Carl I. *Mapping the Trans-Mississippi West, 1540-1861*. 6 vols. San Francisco: Institute of Historical Cartography, 1957-1963.

Wheeler, George Montague. *Preliminary Report Concerning Exploration and Surveys Principally in Nevada and Arizona*. Washington, DC: Government Printing Office, 1872.

Wilford, John Noble. *The Mapmakers*. New York: Alfred A. Knopf, 1981.

Williams, Glyn. *Voyages of Delusion: The Quest for the Northwest Passage*. New Haven: Yale University Press, 2003.

Willson, Marcius. *Outlines of History: Illustrated by Numerous Geographical and Historical Notes and Maps Embracing Part I. Ancient History; Part II. Modern History*. New York: Ivison & Phinney, 1854.

Wood, Denis. *The Power of Maps*. New York: Guilford Press, 1992.

Woodward, David. *The All-American Map: Wax Engraving and Its Influence on Cartography*. Chicago: University of Chicago Press, 1977.

———. *Art and Cartography: Six Historical Essays*. Chicago: University of Chicago Press, 1987.

———. *Five Centuries of Map Printing*. Chicago: University of Chicago Press, 1975.

Yorke, Douglas A., Jr., and John Margolies. *Hitting the Road: The Art of the American Road Map*. San Francisco: Chronicle Books, 1996.

# Cartobibliography

Barreiro, Francisco. *Plano Corográphico e Hydrográphico de las Provincias . . . de la Nueva España*. Manuscript map, 1728. [Hispanic Society of America, New York.]

Beckwith, E. G. *From Great Salt Lake to the Humboldt Mountains. From Explorations and.Surveys Made under the Direction of the Hon. Jefferson Davis, Secretary of War by Capt. E. G. Beckwith 3rd Artillery, E.*[sic] *W. Egloffstein, Topographer for the Route, 1855*. Washington, DC: Selmar Siebert's Engraving & Printing Establishment, 1859. [Library of Congress, Geography and Map Division.]

Bonneville, Benjamin. *Map of the Territory West of the Rocky Mountains*. In U.S. War Dept., *Reports of Explorations and Surveys, to Ascertain the Most Practicable and Economical Route for a Railroad from the Mississippi River to the Pacific Ocean*. Washington, DC: Beverly Tucker, 1855-61. [Special Collections Division, University of Texas at Arlington Libraries, F593 U59 v.11, c.2, p.34.]

Bower, John. *Missouri Territory, Formerly Louisiana*. Philadelphia, 1814. [Special Collections Division, University of Texas at Arlington Libraries, 132/1 #00795.] (See also Lewis, below, for virtually identical map.)

Burr, David H. *Map (1830) by David H. Burr, Geographer to the House of Representatives, Showing the Jedediah Smith Trails*. 1839. [Special Collections Division, University of Texas at Arlington Libraries, 18/4 #800635.]

————. *Map of the United States of North America with Parts of the Adjacent Countries*. By David H. Burr. (Late Topographer to the Post Office.) Geographer to the House of Representatives of the U.S. John Arrowsmith. Entered . . . July 10, 1839, by David H. Burr . . . District of Columbia. [David Rumsey Collection, San Francisco Image ID 140001.]

Cary, John. *A New Map of America from the Latest Authorities.* London: John Cary, Engraver & Map-seller, No. 181, Strand, Decr. 1, 1806. [David Rumsey Collection, San Francisco, Image No. 640038.]

Coronelli, Vincenzo. *America Settentrionale colle Nuove Scoperti all' Anno 1688.* Venice: Coronelli, 1695. [Special Collections Division, University of Texas at Arlington Libraries, 142/7 #00568; 240059.]

Emory, William H. *Map of the United States and Their Territories between the Mississippi and the Pacific Ocean and Part of Mexico Compiled from the Surveys Made under the Order of W. H. Emory. . . . and from the Maps of the Pacific Rail Road, General Land Office, and the Coast Survey.* Projected and drawn under the supervision of Lt. N. Michler, Topl. Engrs. By Thomas Jekyll, C.E. 1857-8. Engraved by Selmar Siebert. Lettering by F. Courtenay. Selmar Siebert's Engraving and Printing Establishment. Washington, DC, 1857. [David Rumsey Collection, San Francisco, Image No. 170002.]

Finley, Anthony. *North America.* In *A New General Atlas, Comprising a Complete Set of Maps Representing the Grand Divisions of the Globe, Together with the Several Empires, Kingdoms and States in the World.* Philadelphia: Anthony Finley, 1830. [Special Collections Division, University of Texas at Arlington Libraries, Oversize G1019 F45 1830.]

Frémont, John Charles. *Map of Oregon and Upper California from the Surveys of John Charles Fremont and Other Authorities.* Drawn by Charles Preuss under the Order of the Senate of the United States, Washington City, 1848. Lithy. by E. Weber & Co. Balto. [inset] Profile of the travelling route from the South Pass of the Rocky Mountains to the Bay of San Francisco. Washington, DC: Wendell and Van Benthuysen, 1848. [David Rumsey Collection, San Francisco, Image No. 170046.]

General Land Office, Dept. of the Interior. *Map of the State of Nevada to Accompany the Annual Report of the Commr. Genl. Land Office, 1866.* New York: Major & Knapp, 1866. [Special Collections Division, University of Texas at Arlington Libraries, 16/11 #800261.]

Goddard, George. *Map of the State of California.* San Francisco: Britton & Rey, 1857. [David Rumsey Collection, San Francisco, Image No. 10036.]

Grässl, J. *Vereinigte Staaten von Nord America: Californien, Texas, und die Territorien New Mexico u. Utah.* Published in Germany by Meyre in 1852. [Special Collections Division, University of Texas at Arlington Libraries, 2000–874.]

Humboldt, Alexander von. *A Map of New Spain from 16° to 38° North Latitude: Reduced from the Large Map Drawn from Astronomical Observations at Mexico in the Year 1804.* London: Longman, Hurst, Rees, Orme & Brown, 1810. [Special

Collections Division, University of Texas at Arlington Libraries, 53/9 #90–1687.]

Lewis, Samuel. *Missouri Territory, Formerly Louisiana*. Philadelphia: M. Carey, 1814. [Special Collections Division, University of Texas at Arlington Libraries, 132/1 #00370.] (See Bower, above.)

Manly, William Lewis. *Map of the Jayhawker's Trail*. Manuscript map, n.d., ca. 1889. [Huntington Library, San Marino, California, HM 50895.]

*Map of the Comstock Lodes Extending down Gold Cañon, Storey Co. Nevada*. San Francisco: G. T. Brown, n.d. [Special Collections Division, University of Texas at Arlington Libraries, 17/11 #800502.]

*Map of the Lower Comstock and Emigrant Consolidated Mining Cos. Mines, Lyon Co*. San Francisco: G. T. Brown & Co., n.d. [Special Collections Division, University of Texas at Arlington Libraries, 16/11 #800501.]

Melish, John. *Map of the United States with the Contiguous British and Spanish Possessions. Compiled by the latest and best Authorities. By John Melish. Engraved by J. Valance and H. S. Tanner. Entered . . . 6th day of June, 1816*. Published by John Melish, Philadelphia, 1816. [David Rumsey Collection, San Francisco, Image No. 20060.]

Miera y Pacheco, Bernardo de. *Plano Geographico, de la tierra descubierta . . . del nuevo Mexico*. Manuscript maps, 1776-1778.

———. a) Plano geografico de la tierra descubierta, y demarcada, por $D^n$. Bernardo de Miera y Pacheco al rumbo del Noroeste, y oeste del nuevo Mexico, quien fue en compañia de los $RR^S$. $PP^S$. Fr. Fran.$^{co}$ Atanacio Dominguez, Visitador Comisario y Custodio de esta, y Fr. Silvestre Velez de Escalante, siendo uno del numero de las diez personas que acompañaron à dichas RR. $PP^S$. como constara en le Diario Derrotero Que hicieron à que se remite en togo con le fin unico del servicio de ambas Magestades, le que va adjunto en dho Diario al Exmo Sor $B^o$. $D^n$ Fray Antonio Maria Bucareli y Ursua Teniente Gral de los $R^S$. Exercitos de S. M., Virrey Govern$^{or}$. Y Capitan Grâl. De estra Nueva España, a quien humilde, y rendidamente dedican esta pequeña Obra por la direccion del coronl. $D^n$. Pedro fermin de Mendinueta Govern$^{or}$. Y Capitan Grâl de este Reno para los fines que puedan conducir al bien de tantas Almas y al servicio de N. Catolico Soverano. Manuscript 32-1/4 by 27-3/4 inches. *[According to cartographic historian Carl Wheat, this is British Museum Additional Manuscript 17,661-C, of which photostats are filed in the Library of Congress and the Newberry Library. This form of Miera's map of the Escalante expedition has been designated "Type A" in the accompanying text by Wheat.]*

———. b) Plano geografico, de la Tierra descubierta, y demarcada por $D^n$.

Bernardo de Miera, y Pacheco al rumbo del Noroeste y oeste, del Nuevo Mexico, quien fue en compania de los RR$^S$. PP$^S$. Fr fran$^{CO}$. Antanacio Domingues, Visitador, Comisario y Custodio de esta ; y fr. Silvestre Velez de Escalante ; siendo uno del numero de las diez personas que acompañaron à dhos RR$^S$. Padres, como constara en el Diario Derrotero que hicieron a que se remite en todo ; con el fin unico del servicio de ambas Magestades el que va adjunto en dicho Diario al Comandante General de las Provincias Internas, el S$^{or}$. Brigadier de los R$^S$. Exercitos Cavaleero Crois, a queien humilde y rendidamente dedican esta pequeña obra, por la Direccion del Coronel D$^n$. Pedro fermin de Mendinueta Governador de esta Reyno, para los fines que puedan conducir al bien de tantas Almas que desean ser Christianas, y al servicio de Nuestro Catholico Soberano. Manuscript, 31-3/4 by 27-3/4 inches. *[Wheat notes that this is in the Archivo de Mapas of the Ministry of War at Madrid (L.M. 8a-1a-a). It is what Wheat calls "Type B," the "Tree and Serpent" type of Miera's map of the Escalante expedition route.]*

————. c) Plano Geographico, de la tierra descubierta, nuevamente, à los Rumbos Norte Noroeste y Oeste, del Nuevo Mexico. Demarcada por mi Don Bernardo de Miera y Pacheco que entrò a hacer su descubrimiento, en compañia de los RR$^S$. PP$^S$. Fr. Francisco Atanacio Doming$^S$. Y Fr. Silbestre Veles, segun constra en el Diario y Derrotero que se hizo y se remitio à S.M. por mano de su Virrei con Plano à la letra: El que dedica Al S$^r$. D. Teodoro de la Crois, del Insigne Orden Teutonica Comandante General en Gefe de Linea y Provincias de esta America Setemptrional, por S. M. heco en S. Ph. El Real de Chiguagua. Ano de 1778. Manuscript 31-7/8 by 27-3/4 inches. Two colored vignettes. *[According to Wheat, this is British Museum Additional Manuscript 17,661-D, and photostats are on file in the Library of Congress and the Newberry Library. Wheat notes that its style differs from others, and that it has been termed the "Bearded Indian."]*

————. d) Same title as 1778-*Miera*, I. Manuscript, 30-7/8 by 25-5/8 inches. *[According to Wheat, this copy is preserved in the Archivo General at Mexico City, and a hand-drawn copy, with some corrections and additions, was reproduced in Herbert E. Bolton,* Pageant in the Wilderness (Salt Lake City: Utah State Historical Society, 1950).]

————. e) Same title as 1778-*Miera*, I. Manuscript, 32 by 25 inches. *[Wheat states that this tracing is in the Kohl Collection at the Library of Congress, No. 271.]*

————. f) Same title as 1778-*Miera*, I. Manuscript, 33-1/4 by 25-3/4 inches. *[According to Wheat, this tracing is also in the Kohl Collection at the Library of Congress, and is apparently numbered 271, like the preceding map, from which it varies slightly.]*

Münster, Sebastian. *Tabula Novarum Insularum, quas Diversis Respectibus Occidentalis & Indianas Vocant.* Basel: Heinrich Petri, 1540. [Special Collections Division, University of Texas at Arlington Libraries, 128/12 #00566.]

Ortelius, Abraham. *Americae sive Novi Orbis, Nova Descriptio.* Antwerp: Aegid, 1570. [Special Collections Division, University of Texas at Arlington Libraries, 24/4 #220022.]

Pike, Zebulon Montgomery. *A Map of the Internal Provinces of New Spain.* Philadelphia, 1810. [Special Collections Division, University of Texas at Arlington Libraries, 27/9 #310162.]

Preuss, Charles. *Map of Oregon and Upper California from the Surveys of John Charles Frémont.* Baltimore: E. Weber., 1848. [Special Collections Division, University of Texas at Arlington Libraries, 77/13 #310103.]

Ravenstein, E. G. *Map of the Southwestern Portion of the United States and Sonora and Chihuahua.* From William A. Bell, *New Tracks in North America: A Journal of Travel and Adventure while Engaged in the Survey for a Southern Railroad to the Pacific Ocean during 1867-8.* London: Chapman and Hall; New York: Scribner, Welford & Co., 1869, vol. 1, betw. pp. xiv and xv. [DeGolyer Library, Southern Methodist University, Collection No. F786.B43.]

Robinson, John. *Map of Mexico, Louisiana, and the Missouri Territory: Including also the State of Mississippi, Alabama Territory, East and West Florida, Georgia, South Carolina & part of the Island of Cuba.* Philadelphia: J. L. Narstin, 1819. [Special Collections Division, University of Texas at Arlington Libraries, 97/1 #3327—framed in VGCHL.]

Ruscelli, Girolamo. *Nueva Hispania Tabula Nova.* Venice: Appresso Giordano Ziletti, ca. 1565. [Special Collections Division, University of Texas at Arlington Libraries, 128/12 #00384.]

Simpson, J. H. *Profiles of Wagon Routes in the Territory of Utah.* In *Report of Explorations across the Great Basin of the Territory of Utah for a Direct Wagon-Route from Camp Floyd to Genoa, in Carson Valley, in 1859.* Washington, DC: Government Printing Office, 1876. [Special Collections Division, University of Texas at Arlington Libraries, Uncataloged #2003–767.]

Smythe, R. M. *Map of Bodie Mining District, Mono Co. Cal.* New York: The Mining Record, 1879. [Special Collections Division, University of Texas at Arlington Libraries, 100/2 #180002.]

Society for the Diffusion of Useful Knowledge. *Central America II, Including Texas, California, and the Northern States of Mexico.* London: Chas. Knight & Co., 1842. [Special Collections Division, University of Texas at Arlington Libraries, 126/12 #00327.]

———. *Central America II, Including Texas, California, and the Northern States of*

*Mexico*. London: Chas. Knight & Co., 1846. [Special Collections Division, University of Texas at Arlington Libraries, 126/12 #00328.]

*Striking Comparison, A.* [Map comparing Utah/Deseret to the Holy Land.] In *Rio Grande Western Railway, Pointer to Prosperity: A Few Facts About the Climate and Resources of the New State of Utah.* Salt Lake City: Rio Grande Western Railway, 1896. [Special Collections, Degolyer Library, Southern Methodist University.]

Walker, Juan Pedro. Map of Western and Central North America. Untitled manuscript map, n.d., ca. 1817. [Huntington Library, San Marino, CA, HM2048.]

Warren, G. K. *Map of the Territory of the United States from the Mississippi River to the Pacific Ocean. . . . To Accompany the Reports of the Explorations for a Railroad Route. . . . Compiled from Authorized Explorations and Other Reliable Data by Lieut G. K. Warren.* By E. Freyhold. New York: J. Bien, 1858. [Special Collections Division, University of Texas at Arlington Libraries, 700004; western half of map only.]

Wheeler, George. *Map Showing Detailed Topography of the Country Traversed by the Reconnaissance Expedition through Southern & Southeastern Nevada in Charge of Lieut. Geo. M. Wheeler, U.S. Engineers. Assisted by O. W. Lockwood, Corps of Engineers U.S.A. P W. Hamel Chief Topographer and Draughtsman.* New York: New York Lithography, Engraving & Printing, 1869. [David Rumsey Collection, San Francisco, Image No. 40024.]

Williams, H. T. *Williams' New Trans-Continental Map of the Pacific R.R. and Routes of Overland Travel to Colorado, Nebraska, the Black Hills, Utah, Idaho, Nevada, Montana, California, and the Pacific Coast.* New York: Adams & Bishop, 1877. [Special Collections Division, University of Texas at Arlington Libraries, 102/8 #800650.]

Willson, Marcius. *The United States and Their Territories.* From *Outlines of History: Illustrated by Numerous Geographical and Historical Notes and Maps Embracing Part I. Ancient History; Part II. Modern History.* New York: Ivison & Phinney; Chicago: S.C. Griggs & Co., 1854, p. xviii. [Fondren Library, Southern Methodist University, Call No. 902 W742.]

Winterbotham, William. *A General Map of North America Drawn from the Best Surveys, 1795.* In *The American Atlas.* New York: John Reid, 1795. [Special Collections Division, University of Texas at Arlington Libraries, Oversize #92–2130.]

# Index